SpringerBriefs in Mathematics

SpringerBriefs in Mathematics showcases expositions in all areas of mathematics and applied mathematics. Manuscripts presenting new results or a single new result in a classical field, new field, or an emerging topic, applications, or bridges between new results and already published works, are encouraged. The series is intended for mathematicians and applied mathematicians.

More information about this series at http://www.springer.com/series/10030

Shukai Du · Francisco-Javier Sayas

An Invitation to the Theory of the Hybridizable Discontinuous Galerkin Method

Projections, Estimates, Tools

Springer

Shukai Du
Mathematical Sciences
University of Delaware
Newark, DE, USA

Francisco-Javier Sayas
Mathematical Sciences
University of Delaware
Newark, DE, USA

ISSN 2191-8198 ISSN 2191-8201 (electronic)
SpringerBriefs in Mathematics
ISBN 978-3-030-27229-6 ISBN 978-3-030-27230-2 (eBook)
https://doi.org/10.1007/978-3-030-27230-2

Mathematics Subject Classification (2010): 65N30, 65N12, 65N15, 65M60

This Springer imprint is published by the registered company Springer Nature Switzerland AG
The registered company address is: Gewerbestrasse 11, 6330 Cham, Switzerland

To Bernardo Cockburn, for his tireless efforts to support entire generations of numerical analysts, and for being an inspiration

Preface

This monograph gives a self-contained presentation of the analysis of the Hybridizable Discontinuous Galerkin method, for some linear elliptic and evolutionary equations.

An early version of this monograph covering most of Chaps. 1–3, and distributed via arXiv under the title *From Raviart–Thomas to HDG: a personal voyage* (the last part of the title was a tribute to Carl Sagan's *Cosmos*) was written by the second author as support material for a summer course as part of *Cádiz Numérica 2013—Course and Encounter on Numerical Analysis* (Cádiz, Spain—June 2013). It was an introduction to the techniques for local analysis of classical mixed methods for diffusion problems and how they motivate the Hybridizable Discontinuous Galerkin method. They assume knowledge of basic techniques on Finite Element Analysis, but not of Mixed Finite Element Methods. The notes have now grown to a much more polished text, including not only the original HDG method for diffusive problems but the treatment of the Helmholtz equations, the Lehrenfeld–Schöberl variant of HDG (which we will call HDG+ here), and the use of HDG for evolutionary equations. Much of this material can be found in articles, although we present it here with a different flavor and some novelties. *Let us emphasize that this book is about the theory of the method and does not give hints at implementation, comparison with other methods, or numerical experiments to illustrate the sharpness of the theoretical results.* All of those aspects can easily be found in the literature and including them in this monograph would change the scope and make this a much longer text.

Many estimates (especially those fitting in the general category of scaling arguments) are carried out with almost excruciating detail. All final bounds are given for solutions with maximal regularity. This being an introductory text, no attempt has been made to deal with complicated or anisotropic meshes, or to produce estimates for solutions with very low regularity. We have preferred to give most technical results in a theorem-and-proof format but have kept a more argumentative (while fully rigorous) tone for the main estimates of the three families of methods we will be studying. We have also tried to give some precise references to original sources but have not been extremely thorough in this, partially because of ambiguities (on the part of the authors) as to who was first in some particular instances.

Our approach can be explained in one line: we want to find a mechanical point of view that streamlines the analysis of HDG methods. This is motivated by the fast growth of the literature on HDG, where a slowly cooked presentation is still lacking. It is also triggered by our belief that sometimes analysis is made difficult by closeness to novelties, need of quick publication, and by cumbersome notation that can typically be seen only when the papers are approached by an outsider. In general, we try to offer an eagle's view that renders some arguments of the HDG analysis quite trivial. This does not mean at all that the analysis is made trivial, but that the key ideas are moved to where they belong (the realm of novelty), while the simple developments are displayed in the sheer simplicity of common tricks.

We now point out some novelties in our approach to the analysis of HDG that pervade this monograph. First of all, we slowly motivate the analysis of HDG by studying two famous mixed methods (the Raviart–Thomas and the Brezzi–Douglas–Marini) as hybridizable methods with an attached projection. We have created a little hat-check trick to handle changes to the reference element (they are plain Piola transforms combined in a primal–dual way) in a purely mechanical way. The HDG projection, which was a key to create a new and more comprehensive analysis of the HDG schemes, is analyzed here by a change to the reference element, which makes the entire approximation analysis somewhat simpler. We also offer a novel analysis of the Lehrenfeld–Schöberl HDG method using a projection. This greatly simplifies the analysis of that class of methods by pushing all the effort to the approximation of a projection tailored to the method.

This monograph would have never been possible without many colleagues that have sparked our interest in HDG schemes, starting with the master and commander of the HDG navy, Bernardo Cockburn (from the University of Minnesota), and continuing with Jay Gopalakrishnan (Portland State University), Gabriel Gatica (University of Concepción, Chile), and Salim Meddahi (University of Oviedo, Spain). All of them have been sources of knowledge and inspiration. In different roles (student and boss), we are members of "Team Pancho", an undisguised effort at the University of Delaware to make Numerical Analysis great, mathematically meaningful, practical, and especially fun.

We finally acknowledge the important support of the NSF Division of Mathematical Sciences through the grant DMS-1818867 *Simulation and numerical analysis in elastodynamics*, which has a key HDG component.

Newark, Delaware, USA — Shukai Du
February 2019 — Francisco-Javier Sayas

Contents

Chapter 1
Getting Ready

This chapter introduces a collection of transformation techniques, equalities, and bounds to relate quantities defined in physical variables to quantities in a reference configuration. They belong to the category of what the FEM community calls *scaling arguments*. We will not be using anything specific from FEM theory, but the arguments will be familiar to anyone aware of these techniques. Proper introductions can be found in classic books as Ciarlet's [24], Brenner and Scott's [6] or Braess's [5].

1.1 A Personal View of Piola Transforms

Reference configurations. Let $\widehat{K}$ be the reference triangle/tetrahedron

$$\widehat{K} := \{\widehat{\mathbf{x}} \in \mathbb{R}^d \ : \widehat{x}_i \geq 0 \quad \forall i, \quad \mathbf{e} \cdot \widehat{\mathbf{x}} \leq 1\}, \quad \text{where } \mathbf{e} = (1, \ldots, 1)^\top, \quad d = 2, 3.$$

Given a general triangle/tetrahedron, we consider fixed affine invertible maps

$$\mathrm{F} : \widehat{K} \to K, \qquad \mathrm{G} := \mathrm{F}^{-1} : K \to \widehat{K},$$

and will denote

$$\mathrm{B} := \mathrm{DF}, \qquad \mathrm{B}^{-1} = \mathrm{DG}, \qquad |J| := |\det \mathrm{B}|,$$

where the differential operator D is defined such that $(\mathrm{DF})_{ji} = \partial_{\widehat{x}_i} \mathrm{F}_j$. At the present moment, it is not necessary to show dependence on K of all of these quantities.

S. Du and F.-J. Sayas, *An Invitation to the Theory of the Hybridizable Discontinuous Galerkin Method*, SpringerBriefs in Mathematics,
https://doi.org/10.1007/978-3-030-27230-2_1

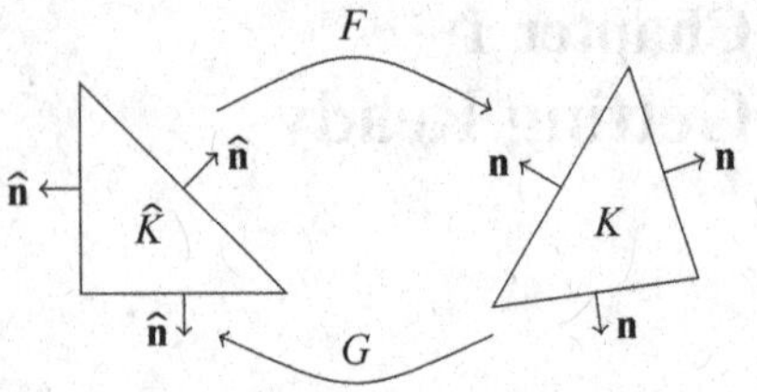

Fig. 1.1 Affine maps F, G between reference element $\widehat{K}$ and physical element K

We will also consider a piecewise constant function $|a| : \partial\widehat{K} \to \mathbb{R}$ containing the absolute value of the determinant of the tangential derivative matrix of $\mathrm{F}|_{\partial\widehat{K}}$. In particular

$$\int_K f = \int_{\widehat{K}} f \circ \mathrm{F}\,|J|, \qquad \int_{\partial K} f = \int_{\partial\widehat{K}} f \circ \mathrm{F}|_{\partial\widehat{K}}\,|a|.$$

Note that the latter equality is not a typical application of the change of variable formula (that applies on volumes) but the possibility of parametrizing both surface integrals from the same sets of coordinates. Outward pointing unit normal fields will be denoted $\mathbf{n} : \partial K \to \mathbb{R}^d$ and $\widehat{\mathbf{n}} : \partial\widehat{K} \to \mathbb{R}^d$.

Changes of variables. We will deal with three types of fields: scalar fields u defined on the volume, vector fields $\mathbf{q}$ defined on the volume, and scalar fields μ defined on the boundary. For reasons we will see repeatedly, inner products are often better understood as duality products. In light of this, we will write

$$(u, u^*)_K := \int_K u\,u^*, \qquad (\mathbf{q}, \mathbf{q}^*)_K := \int_K \mathbf{q}\cdot\mathbf{q}^*, \qquad \langle\mu, \mu^*\rangle_{\partial K} := \int_{\partial K} \mu\,\mu^*,$$

thinking momentarily that starred quantities are dual variables and unstarred quantities are primal. The rules for changes of variables are as follows

Primal	$u : K \to \mathbb{R},$	$\widehat{u} := u \circ \mathrm{F},$
	$\mathbf{q} : K \to \mathbb{R}^d,$	$\widehat{\mathbf{q}} := \lvert J\rvert \mathrm{B}^{-1}\mathbf{q} \circ \mathrm{F},$
	$\mu : \partial K \to \mathbb{R},$	$\widehat{\mu} := \mu \circ \mathrm{F}\lvert_{\partial\widehat{K}},$
Dual	$u^* : K \to \mathbb{R},$	$\check{u}^* := \lvert J\rvert\, u^* \circ \mathrm{F},$
	$\mathbf{q}^* : K \to \mathbb{R}^d,$	$\check{\mathbf{q}}^* := \mathrm{B}^\top \mathbf{q}^* \circ \mathrm{F},$
	$\mu^* : \partial K \to \mathbb{R},$	$\check{\mu}^* := \lvert a\rvert\, \mu^* \circ \mathrm{F}\lvert_{\partial\widehat{K}}.$

so that we can change variables in inner/duality products

$$(u, u^*)_K = (\widehat{u}, \check{u}^*)_{\widehat{K}}, \tag{1.1a}$$

$$(\mathbf{q}, \mathbf{q}^*)_K = (\widehat{\mathbf{q}}, \check{\mathbf{q}}^*)_{\widehat{K}}, \tag{1.1b}$$

$$\langle\mu, \mu^*\rangle_{\partial K} = \langle\widehat{\mu}, \check{\mu}^*\rangle_{\partial\widehat{K}}. \tag{1.1c}$$

This set of rules might be somewhat whimsical, but there are many reasons for them. The best (and deepest) explanations go through p-forms, a context where this and much more make complete sense. The interested reader might want to have a look at the massive work of Arnold et al. [3, 4], where all of this (and considerably more) is given a very general treatment.

A remark on restrictions. The restriction to ∂K of a function $u : K \to \mathbb{R}$ (be it a trace restriction or a more classical one) will be denoted either $u|_{\partial K}$ or just u. Normal components of vector fields $\mathbf{q} : K \to \mathbb{R}^d$, will be denoted $\mathbf{q} \cdot \mathbf{n}$. Note that

$$\widehat{u|_{\partial K}} = \widehat{u}|_{\partial \widehat{K}}, \tag{1.2}$$

a property that is not satisfied by the check transformations (see the definitions of $\check{u}^*$ and $\check{\mu}^*$).

Changes of variables and operators. The following result shows how the gradient and divergence operators, and the normal trace to the boundary transform primal quantities to dual quantities.

Proposition 1.1 (Changes of variables) *For smooth enough fields,*

$$\widehat{\operatorname{div}}\, \widehat{\mathbf{q}} = \widetilde{\operatorname{div} \mathbf{q}}, \tag{1.3a}$$

$$\widehat{\nabla}\, \widehat{u} = \widecheck{\nabla u}, \tag{1.3b}$$

$$\widehat{\mathbf{q}} \cdot \widehat{\mathbf{n}} = \widetilde{\mathbf{q} \cdot \mathbf{n}}, \tag{1.3c}$$

and therefore

$$(\operatorname{div} \mathbf{q}, u)_K = (\widehat{\operatorname{div}}\, \widehat{\mathbf{q}}, \widehat{u})_{\widehat{K}}, \tag{1.4a}$$

$$(\mathbf{q}, \nabla u)_K = (\widehat{\mathbf{q}}, \widehat{\nabla} \widehat{u})_{\widehat{K}}, \tag{1.4b}$$

$$\langle \mathbf{q} \cdot \mathbf{n}, \mu \rangle_{\partial K} = \langle \widehat{\mathbf{q}} \cdot \widehat{\mathbf{n}}, \widehat{\mu} \rangle_{\partial \widehat{K}}. \tag{1.4c}$$

Proof This moment is as good as any other to learn how to use Einstein notation: repeated subindices will denote addition in that index, and a comma followed by one or more indices denotes differentiation w.r.t. the corresponding variable. Note that $(\mathrm{DF})_{ij} = \mathrm{B}_{ij} = \mathrm{F}_{i,j}$. Differentiating in $\widehat{u} = u \circ \mathrm{F}$, we have

$$\widehat{u}_{,i} = (u_{,j} \circ \mathrm{F})\, \mathrm{F}_{j,i} = \mathrm{B}_{ji} u_{,j} \circ \mathrm{F},$$

which is (1.3b). All other formulas can be proved using duality arguments. We will do some of them by hand. For instance, the formula

$$\mathbf{q} = |J|^{-1} \mathrm{B} \widehat{\mathbf{q}} \circ \mathrm{G} \qquad \text{(recall that } \mathrm{G} = \mathrm{F}^{-1})$$

is written componentwise as $q_i = |J|^{-1}\mathrm{B}_{ij}\widehat{q}_j \circ \mathrm{G}$, and leads to

$$\begin{aligned} q_{i,i} &= |J|^{-1}\mathrm{B}_{ij}(\widehat{q}_{j,k} \circ \mathrm{G})\mathrm{G}_{k,i} && \text{(chain rule)} \\ &= |J|^{-1}\mathrm{B}_{ij}\mathrm{B}^{-1}_{ki}\widehat{q}_{j,k} \circ \mathrm{G} && (\mathrm{DG} = \mathrm{B}^{-1}) \\ &= |J|^{-1}\delta_{kj}\widehat{q}_{j,k} \circ \mathrm{G} && (\mathrm{B}^{-1}\mathrm{B} = \mathrm{I}) \\ &= |J|^{-1}\widehat{q}_{j,j} \circ \mathrm{G}, \end{aligned}$$

that is, $|J| q_{i,i} \circ \mathrm{F} = \widehat{q}_{i,i}$, which proves (1.3a). The changes of variables (1.3a) and (1.3b) and the integral rules (1.1)—which motivated our notation—imply (1.4a) and (1.4b). The divergence theorem proves (1.4c). It then follows from (1.1) that

$$\langle \widetilde{\mathbf{q} \cdot \mathbf{n}} - \widehat{\mathbf{q}} \cdot \widehat{\mathbf{n}}, \widehat{\mu} \rangle_{\partial \widehat{K}} = 0.$$

Taking $\widehat{\mu} = \widetilde{\mathbf{q} \cdot \mathbf{n}} - \widehat{\mathbf{q}} \cdot \widehat{\mathbf{n}}$, (1.3c) follows.

1.2 Scaling Inequalities

Some no-brainers. We start with quite obvious changes of variables for integrals

$$\|u\|_K \le |J|^{1/2}\, \|\widehat{u}\|_{\widehat{K}}, \qquad \|\widehat{u}\|_{\widehat{K}} \le |J|^{-1/2}\|u\|_K, \qquad \text{(obviously equal)} \tag{1.5a}$$

$$\|\mathbf{q}\|_K \le |J|^{-1/2}\|\mathrm{B}\|\, \|\widehat{\mathbf{q}}\|_{\widehat{K}}, \qquad \|\widehat{\mathbf{q}}\|_{\widehat{K}} \le |J|^{1/2}\|\mathrm{B}^{-1}\|\, \|\mathbf{q}\|_K, \tag{1.5b}$$

$$\|\mu\|_{\partial K} \le \|a\|^{1/2}_{L^\infty}\|\widehat{\mu}\|_{\partial \widehat{K}}, \qquad \|\widehat{\mu}\|_{\partial \widehat{K}} \le \|a^{-1}\|^{1/2}_{L^\infty}\|\mu\|_{\partial K}. \tag{1.5c}$$

At this precise point, we start assuming that there is a finite collection of triangles/tetrahedra $\mathscr{T}_h$. The diameter of K is denoted h_K. We typically write $h := \max_{K \in \mathscr{T}_h} h_K$. The collection $\mathscr{T}_h$ is called shape-regular when $h_K \le C\rho_K$, where ρ_K is the diameter of the largest ball that we can insert in K. This definition includes a constant $C = C(\mathscr{T}_h)$ that always exists. For it to make sense *with C independent of h*, we have to assume that there is actually a collection of triangulations $\mathscr{T}_h$, that are just tagged with this general parameter h. Readers are supposed to be in the know of this FEM abuse of notation, and we will not insist on this any longer. Wiggled inequalities will be extremely useful to avoid the introduction of constants that are independent of h, possibly different in each occurence:

$$a_h \lesssim b_h \qquad \text{means} \qquad a_h \le C\, b_h \text{ with } C > 0 \text{ independent of } h,$$

and

$$a_h \approx b_h \qquad \text{means} \qquad a_h \lesssim b_h \lesssim a_h.$$

Shape-regularity implies

$$\|\mathrm{B}_K\| \lesssim h_K, \qquad \|\mathrm{B}_K^{-1}\| \lesssim h_K^{-1}, \tag{1.6a}$$
$$|J_K| \lesssim h_K^d, \qquad |J_K|^{-1} \lesssim h_K^{-d}, \qquad (|J_K| \approx h_K^d) \tag{1.6b}$$
$$\|a_K\|_{L^\infty} \lesssim h_K^{d-1}, \qquad \|a_K^{-1}\|_{L^\infty} \lesssim h_K^{1-d}, \tag{1.6c}$$

and then (1.5) can be written as

$$\|u\|_K \approx h_K^{\frac{d}{2}} \|\widehat{u}\|_{\widehat{K}}, \qquad \|\mathbf{q}\|_K \approx h_K^{1-\frac{d}{2}} \|\widehat{\mathbf{q}}\|_{\widehat{K}}, \qquad \|\mu\|_{\partial K} \approx h_K^{\frac{d-1}{2}} \|\widehat{\mu}\|_{\partial\widehat{K}}. \tag{1.7}$$

Sobolev seminorms. When derivatives are introduced (through Sobolev seminorms), the well-known scaling properties for scalar volume fields are

$$|u|_{m,K} \lesssim |J|^{1/2} \|\mathrm{B}^{-1}\|^m |\widehat{u}|_{m,\widehat{K}}, \tag{1.8a}$$
$$|\widehat{u}|_{m,\widehat{K}} \lesssim |J|^{-1/2} \|\mathrm{B}\|^m |u|_{m,K}. \tag{1.8b}$$

Applying these inequalities to the components of $\widehat{\mathbf{q}} \circ \mathrm{G}$, we can prove

$$|\mathbf{q}|_{m,K} \lesssim |J|^{-1/2} \|\mathrm{B}\|\, \|\mathrm{B}^{-1}\|^m |\widehat{\mathbf{q}}|_{m,\widehat{K}}, \tag{1.8c}$$
$$|\widehat{\mathbf{q}}|_{m,\widehat{K}} \lesssim |J|^{1/2} \|\mathrm{B}^{-1}\|\, \|\mathrm{B}\|^m |\mathbf{q}|_{m,K}. \tag{1.8d}$$

This and shape-regularity (1.6) yield

$$|u|_{m,K} \approx h_K^{\frac{d}{2}-m} |\widehat{u}|_{m,\widehat{K}}, \qquad |\mathbf{q}|_{m,K} \approx h_K^{1-\frac{d}{2}-m} |\widehat{\mathbf{q}}|_{m,\widehat{K}}. \tag{1.9}$$

Exercises

1. Prove the local trace inequality

$$h_K^{1/2} \|v\|_{\partial K} \lesssim \|v\|_K + h_K \|\nabla v\|_K \qquad \forall v \in H^1(K).$$

Chapter 2
Projection Analysis of Mixed Methods

2.1 The Raviart–Thomas Projection

In this section, we will review some well-known (and some not so well-known) facts about the natural interpolation operator associated to the Raviart–Thomas space. Their original and very often quoted paper [109] contains a two-dimensional finite element for the $\mathbf{H}(\mathrm{div}, \Omega)$ space, which is slightly different from the one that is now known as the RT space. The three-dimensional space is one of the many elements that appears in the first of the two big finite element papers by Nédélec [92, 93].

2.1.1 *Facts You Might (not) Know About Polynomials*

Polynomials. Polynomials in d variables with (total) degree at most k will be denoted $\mathscr{P}_k$. It is often convenient to recall the dimension by reminding the reader where the polynomials are defined. To avoid being too wordy, here is some fast notation:

- $\mathscr{P}_k(K)$, where $K \in \mathscr{T}_h$, is the space of polynomials of degree at most k defined on the element K.
- Whenever needed, we will just write $\mathscr{P}_{-1}(K) = \{0\}$, to avoid singling out some particular cases.
- $\boldsymbol{\mathscr{P}}_k(K) := \mathscr{P}_k(K)^d$.
- $\widetilde{\mathscr{P}}_k(K)$ will denote the space of homogeneous polynomials of degree k.
- $\mathbf{m} \in \mathscr{P}_1(K)^d$ is the function $\mathbf{m}(\mathbf{x}) := \mathbf{x}$. There is a tradition calling this function just $\mathbf{x}$, but then $\widehat{\mathbf{x}}$ has two possible meanings, one as the variable in the reference element, and the other one as the function.

$$\widehat{\mathbf{m}}(\widehat{\mathbf{x}}) = |J|(\widehat{\mathbf{x}} + \mathrm{B}^{-1}\mathbf{b}), \qquad \text{where } \mathbf{b} = \mathrm{F}(\mathbf{0}). \tag{2.1}$$

S. Du and F.-J. Sayas, *An Invitation to the Theory of the Hybridizable Discontinuous Galerkin Method*, SpringerBriefs in Mathematics,
https://doi.org/10.1007/978-3-030-27230-2_2

- $\mathcal{E}(K)$ is the set of edges of the triangle K or faces of the tetrahedron K (so that $\cup_{e\in\mathcal{E}(K)}\overline{e} = \partial K$). We will simply call everything a *face*, while using the letter e (as in edge) to refer to these edges/faces.
- $\mathcal{P}_k(e)$ with $e \in \mathcal{E}(K)$ is the space of $(d-1)$-variate polynomials in tangential coordinates.
- $\mathcal{R}_k(\partial K) = \prod_{e\in\mathcal{E}(K)} \mathcal{P}_k(e)$ are piecewise polynomial functions on ∂K.

Easy facts about dimensions:

$$\dim \mathcal{P}_k(K) = \binom{k+d}{d}, \qquad \dim \widetilde{\mathcal{P}}_k(K) = \dim \mathcal{P}_k(e) = \binom{k+d-1}{d-1},$$

$$\dim \mathcal{R}_k(\partial K) = (d+1)\binom{k+d-1}{d-1}.$$

Two more spaces we will use are

$$\mathcal{P}_k^\perp(K) := \{u \in \mathcal{P}_k(K) : (u, v)_K = 0 \quad \forall v \in \mathcal{P}_{k-1}(K)\},$$
$$\boldsymbol{\mathcal{P}}_k^\perp(K) := \mathcal{P}_k^\perp(K)^d = \{\mathbf{q} \in \boldsymbol{\mathcal{P}}_k(K) : (\mathbf{q}, \mathbf{r})_K = 0 \quad \forall \mathbf{r} \in \boldsymbol{\mathcal{P}}_{k-1}(K)\}.$$

The following decompositions are direct orthogonal sums:

$$\mathcal{P}_k(K) = \mathcal{P}_{k-1}(K) \oplus \mathcal{P}_k^\perp(K), \qquad \boldsymbol{\mathcal{P}}_k(K) = \boldsymbol{\mathcal{P}}_{k-1}(K) \oplus \boldsymbol{\mathcal{P}}_k^\perp(K).$$

It is also clear that

$$\dim \mathcal{P}_k^\perp(K) = \dim \widetilde{\mathcal{P}}_k(K) = \dim \mathcal{P}_k(e), \qquad e \in \mathcal{E}(K). \tag{2.2}$$

Lemma 2.1

(a) If $u \in \mathcal{P}_k^\perp(K)$ satisfies $u|_e = 0$ on some $e \in \mathcal{E}(K)$, then $u = 0$.
(b) If $\mathbf{q} \in \boldsymbol{\mathcal{P}}_k^\perp(K)$ satisfies $\mathbf{q}\cdot\mathbf{n} = 0$ on ∂K, then $\mathbf{q} = \mathbf{0}$.

Proof Part (a) is quite simple. The face e is contained in the hyperplane $p(\mathbf{x}) = \mathbf{x}\cdot\mathbf{n}_e - c = 0$, and then $u = p\,v$, where $v \in \mathcal{P}_{k-1}(K)$. But then,

$$0 = (u, v)_K = (p\,v, v)_K = (p, v^2)_K \qquad \text{while } p < 0 \text{ in } K,$$

so $v = 0$. To prove part (b), we use (a) applied to the polynomial $\mathbf{q}\cdot\mathbf{n}_e$ for each $e \in \mathcal{E}(K)$. Then $\mathbf{q}\cdot\mathbf{n}_e = 0$ in K for every $e \in \mathcal{E}(K)$. This shows that $\mathbf{q} = \mathbf{0}$. Note that the result also holds if $\mathbf{q}\cdot\mathbf{n} = 0$ on $\partial K \setminus e$, for any $e \in \mathcal{E}(K)$, since only d normal vectors are needed to have a basis of $\mathbb{R}^d$.

Lemma 2.2 *The following decomposition is a direct orthogonal sum:*

$$\mathscr{R}_k(\partial K) = \{u|_{\partial K} : u \in \mathscr{P}_k^{\perp}(K)\} \oplus \{\mathbf{q} \cdot \mathbf{n} : \mathbf{q} \in \boldsymbol{\mathscr{P}}_k^{\perp}(K)\}. \tag{2.3}$$

Proof By Lemma 2.1, the operators $R_1 : \mathscr{P}_k^{\perp}(K) \to \mathscr{R}_k(\partial K)$ and $R_2 : \boldsymbol{\mathscr{P}}_k^{\perp}(K) \to \mathscr{R}_k(\partial K)$, given by

$$R_1 u := u|_{\partial K}, \qquad R_2 \mathbf{q} := \mathbf{q} \cdot \mathbf{n},$$

are one-to-one. On the other hand

$$\langle R_1 u, R_2 \mathbf{q}\rangle_{\partial K} = \langle u, \mathbf{q} \cdot \mathbf{n}\rangle_{\partial K} = (\nabla u, \mathbf{q})_K + (u, \operatorname{div} \mathbf{q})_K = 0,$$

since $\nabla u \in \boldsymbol{\mathscr{P}}_{k-1}(K)$ and $\operatorname{div} \mathbf{q} \in \mathscr{P}_{k-1}(K)$. This means that the sum Range $R_1 \oplus$ Range R_2 (the right-hand side of (2.3)) is orthogonal. The result follows from an easy dimension count:

$$\begin{aligned}
&\dim\,(\text{Range } R_1 \oplus \text{Range } R_2)\\
&= \dim \text{Range } R_1 + \dim \text{Range } R_2 && \text{(direct sum)}\\
&= \dim \mathscr{P}_k^{\perp}(K) + \dim \boldsymbol{\mathscr{P}}_k^{\perp}(K) && (R_1 \text{ and } R_2 \text{ are 1-1})\\
&= (d+1)\dim \mathscr{P}_k^{\perp}(K) = \dim \mathscr{R}_k(\partial K). && \text{(by (2.2))}
\end{aligned}$$

This completes the proof. (This simple lemma appears in [57].)

Polynomials and Piola transforms. It is also easy to note that polynomials are preserved by the changes of variables, in both possible roles of primal and dual functions:

$$\begin{array}{llcl}
\text{Primal} & u \in \mathscr{P}_k(K) & \Longleftrightarrow & \widehat{u} \in \mathscr{P}_k(\widehat{K}),\\
& \mathbf{q} \in \boldsymbol{\mathscr{P}}_k(K) & \Longleftrightarrow & \widehat{\mathbf{q}} \in \boldsymbol{\mathscr{P}}_k(\widehat{K}),\\
& \mu \in \mathscr{R}_k(\partial K) & \Longleftrightarrow & \widehat{\mu} \in \mathscr{R}_k(\partial \widehat{K}),\\
\text{Dual} & u^* \in \mathscr{P}_k(K) & \Longleftrightarrow & \check{u}^* \in \mathscr{P}_k(\widehat{K}),\\
& \mathbf{q}^* \in \boldsymbol{\mathscr{P}}_k(K) & \Longleftrightarrow & \check{\mathbf{q}}^* \in \boldsymbol{\mathscr{P}}_k(\widehat{K}),\\
& \mu^* \in \mathscr{R}_k(\partial K) & \Longleftrightarrow & \check{\mu}^* \in \mathscr{R}_k(\partial \widehat{K}).
\end{array}$$

These relations and (1.1) show that the spaces $\mathscr{P}_k^{\perp}(K)$ and $\boldsymbol{\mathscr{P}}_k^{\perp}(K)$ are also preserved with the changes of variables:

$$\begin{array}{llcl}
\text{Primal} & u \in \mathscr{P}_k^{\perp}(K) & \Longleftrightarrow & \widehat{u} \in \mathscr{P}_k^{\perp}(\widehat{K}),\\
& \mathbf{q} \in \boldsymbol{\mathscr{P}}_k^{\perp}(K) & \Longleftrightarrow & \widehat{\mathbf{q}} \in \boldsymbol{\mathscr{P}}_k^{\perp}(\widehat{K}),\\
\text{Dual} & u^* \in \mathscr{P}_k^{\perp}(K) & \Longleftrightarrow & \check{u}^* \in \mathscr{P}_k^{\perp}(\widehat{K}),\\
& \mathbf{q}^* \in \boldsymbol{\mathscr{P}}_k^{\perp}(K) & \Longleftrightarrow & \check{\mathbf{q}}^* \in \boldsymbol{\mathscr{P}}_k^{\perp}(\widehat{K}).
\end{array}$$

2.1.2 The Space and the Projection

The Raviart–Thomas space. The RT space in K is defined as

$$\mathcal{RT}_k(K) := \boldsymbol{\mathcal{P}}_k(K) \oplus \mathbf{m}\, \widetilde{\mathcal{P}}_k(K) \qquad \text{(recall that } \mathbf{m}(\mathbf{x}) = \mathbf{x}\text{)}.$$

It is quite obvious that

$$\boldsymbol{\mathcal{P}}_k(K) \subset \mathcal{RT}_k(K) \subset \boldsymbol{\mathcal{P}}_{k+1}(K),$$

both inclusions being proper, and

$$\begin{aligned}\dim \mathcal{RT}_k(K) &= d\binom{k+d}{d} + \binom{k+d-1}{d-1}\\ &= \dim \boldsymbol{\mathcal{P}}_{k-1}(K) + \dim \mathcal{R}_k(\partial K).\end{aligned} \tag{2.4}$$

(The last equality takes one minute to prove.) Slightly less obvious facts are collected in the next proposition.

Proposition 2.1

(a) $\mathbf{q}\cdot\mathbf{n} \in \mathcal{R}_k(\partial K)$ *for all* $\mathbf{q} \in \mathcal{RT}_k(K)$.
(b) $\mathbf{q} \in \mathcal{RT}_k(K)$ *if and only if* $\widehat{\mathbf{q}} \in \mathcal{RT}_k(\widehat{K})$.
(c) *If* $\operatorname{div}\mathbf{q} = 0$ *with* $\mathbf{q} \in \mathcal{RT}_k(K)$, *then* $\mathbf{q} \in \boldsymbol{\mathcal{P}}_k(K)$.
(d) $\operatorname{div}\mathcal{RT}_k(K) = \mathcal{P}_k(K)$.

Proof It is clear that to prove (a)-(b) we only need to worry about functions $\mathbf{m}\,p$, where $p \in \widetilde{\mathcal{P}}_k(K)$. It is also clear that $\mathbf{m}\cdot\mathbf{n} \in \mathcal{R}_0(\partial K)$ (the faces are parts of planes with normal vector $\mathbf{n}$, so $\mathbf{x}\cdot\mathbf{n} = c$). Then $(\mathbf{m}\,p)|_{\partial K}\cdot\mathbf{n} \in \mathcal{R}_0(\partial K)\cdot\mathcal{R}_k(\partial K) \subset \mathcal{R}_k(\partial K)$, which proves (a). Part (b) follows from (2.1), that is from the fact that $\widehat{\mathbf{m}}(\widehat{\mathbf{x}}) = |J|\widehat{\mathbf{x}} + \mathbf{c}$, where $\mathbf{c} := |J|\mathrm{B}^{-1}\mathbf{b}$.

If $\mathbf{q} = \mathbf{p} + \mathbf{m}\,p$, with $\mathbf{p} \in \boldsymbol{\mathcal{P}}_k(K)$ and $p \in \widetilde{\mathcal{P}}_k(K)$, then by Euler's homogeneous function theorem:

$$\begin{aligned}\operatorname{div}(\mathbf{p} + \mathbf{m}\,p) &= \operatorname{div}\mathbf{p} + \mathbf{m}\cdot\nabla p + (\operatorname{div}\mathbf{m})\,p\\ &= \operatorname{div}\mathbf{p} + (k+d)\,p \in \mathcal{P}_{k-1}(K) \oplus \widetilde{\mathcal{P}}_k(K),\end{aligned} \tag{2.5}$$

and therefore $p = 0$ if $\operatorname{div}\mathbf{q} = 0$. This proves (c).

Since $\mathcal{RT}_k(K) \subset \boldsymbol{\mathcal{P}}_{k+1}(K)$, it is obvious that $\operatorname{div}\mathcal{RT}_k(K) \subseteq \mathcal{P}_k(K)$. Now, given $u \in \mathcal{P}_k(K)$, we write

$$u = u_0 + u_1 + \cdots + u_k, \qquad u_j \in \widetilde{\mathcal{P}}_j(K) \quad \forall j,$$

and then use Euler's homogeneous function theorem and the computation in (2.5) to obtain

$$\mathbf{p} = \left(\sum_{j=0}^{k} \tfrac{1}{j+d} u_j \right) \mathbf{m} \in \mathbf{m}\mathscr{P}_k(K) \subset \mathscr{RT}_k(K).$$

A simple computation then shows that $\operatorname{div} \mathbf{p} = u$.

The Raviart–Thomas projection. Let $\mathbf{q} : K \to \mathbb{R}^d$ be sufficiently smooth. The RT projection is $\boldsymbol{\Pi}^{\mathrm{RT}}\mathbf{q} \in \mathscr{RT}_k(K)$, characterized by the equations

$$(\boldsymbol{\Pi}^{\mathrm{RT}}\mathbf{q}, \mathbf{r})_K = (\mathbf{q}, \mathbf{r})_K \qquad \forall \mathbf{r} \in \boldsymbol{\mathscr{P}}_{k-1}(K), \tag{2.6a}$$

$$\langle \boldsymbol{\Pi}^{\mathrm{RT}}\mathbf{q} \cdot \mathbf{n}, \mu \rangle_{\partial K} = \langle \mathbf{q} \cdot \mathbf{n}, \mu \rangle_{\partial K} \qquad \forall \mu \in \mathscr{R}_k(\partial K). \tag{2.6b}$$

Attached to this projection, there is a scalar field projection, Π_k, which is just the $L^2(K)$-projection onto $\mathscr{P}_k(K)$:

$$(\Pi_k u, v)_K = (u, v)_K \qquad \forall v \in \mathscr{P}_k(K). \tag{2.6c}$$

Note that as $\mathscr{P}_{-1}(K) = \{0\}$, Eqs. (2.6a) are void for $k = 0$.

Proposition 2.2 (Definition of the RT projection) *Equations (2.6) are uniquely solvable and therefore define a projection onto* $\mathscr{RT}_k(K)$.

Proof Note that (2.4) implies that (2.6) is equivalent to a square system of linear equations, so we only need to prove uniqueness of solution. Let then $\mathbf{q} \in \mathscr{RT}_k(K)$ satisfy

$$(\mathbf{q}, \mathbf{r})_K = 0 \qquad \forall \mathbf{r} \in \boldsymbol{\mathscr{P}}_{k-1}(K), \tag{2.7a}$$

$$\langle \mathbf{q} \cdot \mathbf{n}, \mu \rangle_{\partial K} = 0 \qquad \forall \mu \in \mathscr{R}_k(\partial K). \tag{2.7b}$$

Therefore

$$\|\operatorname{div} \mathbf{q}\|_K^2 = \langle \mathbf{q} \cdot \mathbf{n}, (\operatorname{div} \mathbf{q})|_{\partial K} \rangle_{\partial K} - (\mathbf{q}, \nabla(\operatorname{div} \mathbf{q}))_K = 0$$

by (2.7a) and (2.7b). This implies that $\operatorname{div} \mathbf{q} = 0$ and then, by Proposition 2.1(c), it follows that $\mathbf{q} \in \boldsymbol{\mathscr{P}}_k(K)$. Then (2.7a) means that $\mathbf{q} \in \boldsymbol{\mathscr{P}}_k^{\perp}(K)$, while (2.7b) implies that $\mathbf{q} \cdot \mathbf{n} = 0$. Using Lemma 2.1(b), it follows that $\mathbf{q} = \mathbf{0}$.

The commutativity property. Note that for all $u \in \mathscr{P}_k(K)$,

$$\begin{aligned}(\operatorname{div} \boldsymbol{\Pi}^{\mathrm{RT}}\mathbf{q}, u)_K &= \langle \boldsymbol{\Pi}^{\mathrm{RT}}\mathbf{q} \cdot \mathbf{n}, u \rangle_{\partial K} - (\boldsymbol{\Pi}^{\mathrm{RT}}\mathbf{q}, \nabla u)_K \\ &= \langle \mathbf{q} \cdot \mathbf{n}, u \rangle_{\partial K} - (\mathbf{q}, \nabla u)_K = (\operatorname{div} \mathbf{q}, u)_K,\end{aligned}$$

i.e.,

$$\operatorname{div} \boldsymbol{\Pi}^{\mathrm{RT}}\mathbf{q} = \Pi_k \operatorname{div} \mathbf{q}. \tag{2.8}$$

Invariance by Piola transforms. Our next goal is to relate the RT projection in the physical element (2.6) with the one defined in the reference element: given $\widehat{\mathbf{q}}$ we look for $\widehat{\boldsymbol{\Pi}}^{\mathrm{RT}}\widehat{\mathbf{q}} \in \mathscr{RT}_k(\widehat{K})$, satisfying

$$(\widehat{\boldsymbol{\Pi}}^{\mathrm{RT}}\widehat{\mathbf{q}}, \mathbf{r})_{\widehat{K}} = (\widehat{\mathbf{q}}, \mathbf{r})_{\widehat{K}} \qquad \forall \mathbf{r} \in \mathscr{P}_{k-1}(\widehat{K}), \tag{2.9a}$$

$$\langle \widehat{\boldsymbol{\Pi}}^{\mathrm{RT}}\widehat{\mathbf{q}} \cdot \widehat{\mathbf{n}}, \mu \rangle_{\partial\widehat{K}} = \langle \widehat{\mathbf{q}} \cdot \widehat{\mathbf{n}}, \mu \rangle_{\partial\widehat{K}} \qquad \forall \mu \in \mathscr{R}_k(\partial\widehat{K}). \tag{2.9b}$$

Note that by (1.1)

$$(\widehat{\boldsymbol{\Pi}^{\mathrm{RT}}\mathbf{q}}, \check{\mathbf{r}})_{\widehat{K}} = (\boldsymbol{\Pi}^{\mathrm{RT}}\mathbf{q}, \mathbf{r})_K = (\mathbf{q}, \mathbf{r})_K = (\widehat{\mathbf{q}}, \check{\mathbf{r}})_{\widehat{K}} \qquad \forall \mathbf{r} \in \mathscr{P}_{k-1}(K),$$

and by (1.4c)

$$\begin{aligned}\langle \widehat{\boldsymbol{\Pi}^{\mathrm{RT}}\mathbf{q}} \cdot \widehat{\mathbf{n}}, \check{\mu} \rangle_{\partial\widehat{K}} &= \langle \boldsymbol{\Pi}^{\mathrm{RT}}\mathbf{q} \cdot \mathbf{n}, \mu \rangle_{\partial K} = \langle \mathbf{q} \cdot \mathbf{n}, \mu \rangle_{\partial K} \\ &= \langle \widehat{\mathbf{q}} \cdot \widehat{\mathbf{n}}, \check{\mu} \rangle_{\partial\widehat{K}} \qquad \forall \mu \in \mathscr{R}_k(\partial K).\end{aligned}$$

However, since the test spaces transform well under the check rules and so does the RT space w.r.t. the hat rule (Proposition 2.1(b)), it follows that

$$\widehat{\boldsymbol{\Pi}}^{\mathrm{RT}}\widehat{\mathbf{q}} = \widehat{\boldsymbol{\Pi}^{\mathrm{RT}}\mathbf{q}}. \tag{2.10}$$

2.1.3 Estimates and Liftings

By looking at the equations on the reference element (2.9), and using a basis of the space $\mathscr{RT}_k(\widehat{K})$, it is easy to see that

$$\|\widehat{\boldsymbol{\Pi}}^{\mathrm{RT}}\widehat{\mathbf{q}}\|_{\widehat{K}} \lesssim \|\widehat{\mathbf{q}}\|_{\widehat{K}} + \|\widehat{\mathbf{q}} \cdot \widehat{\mathbf{n}}\|_{\partial\widehat{K}} \lesssim \|\widehat{\mathbf{q}}\|_{1,\widehat{K}} \qquad \forall \widehat{\mathbf{q}} \in \mathbf{H}^1(\widehat{K}) := H^1(\widehat{K})^d. \tag{2.11}$$

This inequality actually shows how the RT projection is well defined on $\mathbf{H}^{\frac{1}{2}+\varepsilon}(\widehat{K})$, which is a space that guarantees the existence of a classical trace operator, so that we have $\widehat{\mathbf{q}} \cdot \widehat{\mathbf{n}} \in L^2(\partial\widehat{K})$. (We will not deal with these low regularity cases in these notes though.) Another easy fact follows from a compactness argument (a.k.a. the Bramble–Hilbert lemma. See [6, Lemma 4.3.8]): since $\widehat{\boldsymbol{\Pi}}^{\mathrm{RT}}$ preserves the space $\mathscr{P}_k(\widehat{K}) \subset \mathscr{RT}_k(\widehat{K})$, then

$$\|\widehat{\mathbf{q}} - \widehat{\boldsymbol{\Pi}}^{\mathrm{RT}}\widehat{\mathbf{q}}\|_{\widehat{K}} \lesssim |\widehat{\mathbf{q}}|_{k+1,\widehat{K}} \qquad \forall \widehat{\mathbf{q}} \in \mathbf{H}^{k+1}(\widehat{K}). \tag{2.12}$$

Proposition 2.3 (Estimates for the RT projection) *On shape-regular triangulations and for sufficiently smooth* $\mathbf{q}$,

(a) $\|\boldsymbol{\Pi}^{\mathrm{RT}}\mathbf{q}\|_K \lesssim \|\mathbf{q}\|_K + h_K|\mathbf{q}|_{1,K}$,

(b) $\|\mathbf{q} - \boldsymbol{\Pi}^{\mathrm{RT}}\mathbf{q}\|_K \lesssim h_K^{k+1}|\mathbf{q}|_{k+1,K}$,
(c) $\|\operatorname{div}\mathbf{q} - \operatorname{div}\boldsymbol{\Pi}^{\mathrm{RT}}\mathbf{q}\|_K \lesssim h_K^{k+1}|\operatorname{div}\mathbf{q}|_{k+1,K}$.

Proof The results follow from the estimates in the reference element (2.11)–(2.12), the relation between the projection and the projection on the reference element (2.10), and scaling arguments (1.7) and (1.9) (or their more primitive forms in (1.5) and (1.8)). For instance

$$
\begin{aligned}
\|\boldsymbol{\Pi}^{\mathrm{RT}}\mathbf{q}\|_K &\leqslant |J_K|^{-1/2}\|\mathrm{B}_K\|\|\widehat{\boldsymbol{\Pi}^{\mathrm{RT}}\mathbf{q}}\|_{\widehat{K}} && \text{(by (1.5))}\\
&= |J_K|^{-1/2}\|\mathrm{B}_K\|\|\widehat{\boldsymbol{\Pi}}^{\mathrm{RT}}\widehat{\mathbf{q}}\|_{\widehat{K}} && \text{(by (2.10))}\\
&\lesssim |J_K|^{-1/2}\|\mathrm{B}_K\|\|\widehat{\mathbf{q}}\|_{1,\widehat{K}} && \text{(by (2.11))}\\
&\lesssim \|\mathrm{B}_K\|\,\|\mathrm{B}_K^{-1}\|\,(\|\mathbf{q}\|_K + \|\mathrm{B}_K\|\,|\mathbf{q}|_{1,K}). && \text{(by (1.5) and (1.8))}
\end{aligned}
$$

Similarly

$$
\begin{aligned}
&\|\mathbf{q} - \boldsymbol{\Pi}^{\mathrm{RT}}\mathbf{q}\|_K\\
&\quad\leqslant |J_K|^{-1/2}\|\mathrm{B}_K\|\,\|\widehat{\mathbf{q}} - \widehat{\boldsymbol{\Pi}^{\mathrm{RT}}\mathbf{q}}\|_{\widehat{K}} && \text{(by (1.5))}\\
&\quad= |J_K|^{-1/2}\|\mathrm{B}_K\|\,\|\widehat{\mathbf{q}} - \widehat{\boldsymbol{\Pi}}^{\mathrm{RT}}\widehat{\mathbf{q}}\|_{\widehat{K}} && \text{(by (2.10))}\\
&\quad\lesssim |J_K|^{-1/2}\|\mathrm{B}_K\|\,|\widehat{\mathbf{q}}|_{k+1,\widehat{K}} && \text{(by (2.12), i.e., Bramble–Hilbert)}\\
&\quad\leqslant \|\mathrm{B}_K^{-1}\|\,\|\mathrm{B}_K\|^{k+2}|\mathbf{q}|_{k+1,K}. && \text{(by (1.8))}
\end{aligned}
$$

To go from these inequalities to (a) and (b), we only need to use the shape-regularity bounds (1.6). To prove (c), we use the commutativity property (2.8) and a bunch of scaling arguments:

$$
\begin{aligned}
&\|\operatorname{div}\mathbf{q} - \operatorname{div}\boldsymbol{\Pi}^{\mathrm{RT}}\mathbf{q}\|_K\\
&\quad= \|\operatorname{div}\mathbf{q} - \Pi_k\operatorname{div}\mathbf{q}\|_K && \text{(by commutativity (2.8))}\\
&\quad\leqslant |J_K|^{1/2}\|\widehat{\operatorname{div}\mathbf{q}} - \widehat{\Pi_k\operatorname{div}\mathbf{q}}\|_{\widehat{K}} && \text{(by (1.5))}\\
&\quad= |J_K|^{1/2}\|\widehat{\operatorname{div}\mathbf{q}} - \widehat{\Pi}_k\widehat{\operatorname{div}\mathbf{q}}\|_{\widehat{K}} && \text{(easy argument)}\\
&\quad\lesssim |J_K|^{1/2}|\widehat{\operatorname{div}\mathbf{q}}|_{k+1,\widehat{K}} && \text{(compactness-Bramble–Hilbert)}\\
&\quad\leqslant \|\mathrm{B}_K\|^{k+1}|\operatorname{div}\mathbf{q}|_{k+1,K}. && \text{(by (1.8))}
\end{aligned}
$$

The result now follows readily.

Proposition 2.4 (RT local lifting of the normal trace) *There exists a linear operator* $\mathbf{L}^{\mathrm{RT}} : \mathscr{R}_k(\partial K) \to \mathscr{RT}_k(K)$ *such that*

$$
(\mathbf{L}^{\mathrm{RT}}\mu)\cdot\mathbf{n} = \mu \quad \textit{and} \quad \|\mathbf{L}^{\mathrm{RT}}\mu\|_K \lesssim h_K^{1/2}\|\mu\|_{\partial K} \qquad \forall \mu \in \mathscr{R}_k(\partial K).
$$

Proof Let $\mathbf{q} = \mathbf{L}^{\mathrm{RT}}\mu \in \mathscr{RT}_k(K)$ be defined as

$$
\mathbf{q} := |J_K|^{-1}\mathrm{B}_K\widehat{\mathbf{q}} \circ \mathrm{G}_K,
$$

where $\widehat{\mathbf{q}} \in \mathscr{RT}_k(\widehat{K})$ is the solution of the discrete equations in the reference domain:

$$(\widehat{\mathbf{q}}, \mathbf{r})_{\widehat{K}} = 0 \qquad \forall \mathbf{r} \in \mathscr{P}_{k-1}(\widehat{K}), \tag{2.13a}$$

$$\langle \widehat{\mathbf{q}} \cdot \widehat{\mathbf{n}}, \xi \rangle_{\partial \widehat{K}} = \langle \check{\mu}, \xi \rangle_{\partial \widehat{K}} \qquad \forall \xi \in \mathscr{R}_k(\partial \widehat{K}). \tag{2.13b}$$

Note that by (1.4c)

$$\langle \mathbf{q} \cdot \mathbf{n}, \xi \rangle_{\partial K} = \langle \widehat{\mathbf{q}} \cdot \widehat{\mathbf{n}}, \widehat{\xi} \rangle_{\partial \widehat{K}} = \langle \check{\mu}, \widehat{\xi} \rangle_{\partial \widehat{K}} = \langle \mu, \xi \rangle_{\partial K} \qquad \forall \xi \in \mathscr{R}_k(\partial K),$$

and therefore $\mathbf{q} \cdot \mathbf{n} = \mu$. Also

$$\begin{aligned} \|\mathbf{q}\|_K &\leqslant |J_K|^{-1/2} \|\mathrm{B}_K\| \|\widehat{\mathbf{q}}\|_{\widehat{K}} && \text{(by (1.5b))} \\ &\lesssim |J_K|^{-1/2} \|\mathrm{B}_K\| \, \|\check{\mu}\|_{\partial \widehat{K}} && \text{(finite dim argument on (2.13))} \\ &\lesssim |J_K|^{-1/2} \|\mathrm{B}_K\| \, \|a\|_{L^\infty}^{1/2} \|\mu\|_{\partial K}, && \text{(simple argument based on (1.5c))} \end{aligned}$$

and the bound follows from estimating all the above geometric quantities using (1.6)

$$\|\mathbf{q}\|_K \lesssim |J_K|^{-1/2} \|\mathrm{B}_K\| \, \|a\|_{L^\infty}^{1/2} \|\mu\|_{\partial K} \lesssim h_K^{-\frac{d}{2}} h_K h_K^{\frac{d-1}{2}} \|\mu\|_{\partial K} = h_K^{1/2} \|\mu\|_{\partial K}.$$

This completes the proof.

More jargon. When in the middle of a Finite Element argument, we use that we are dealing with polynomials of a fixed degree (or any finite-dimensional space) on the reference domain, it is common to refer to the argument as a *finite-dimensional argument*. This leads to inequalities with constants depending on polynomial degrees and dimension but on nothing else.

2.2 Projection-Based Analysis of RT

In this section, we are going to develop a fully detailed analysis of the RT approximation of the system

$$\kappa^{-1}\mathbf{q} + \nabla u = \mathbf{0} \qquad \text{in } \Omega, \tag{2.14a}$$

$$\operatorname{div} \mathbf{q} = f \qquad \text{in } \Omega, \tag{2.14b}$$

$$u = g \qquad \text{on } \Gamma := \partial\Omega, \tag{2.14c}$$

where Ω is a polygonal/polyhedral domain, $f \in L^2(\Omega)$, $\kappa \in L^\infty(\Omega)$ is strictly positive (so that $\kappa^{-1} \in L^\infty(\Omega)$ as well), and $g \in H^{1/2}(\Gamma)$. We are not going to use any of the results of the Brezzi theory of mixed problems [7]. Our approach is going to be more local and much less abstract. (It has to be noted that Brezzi's theory and Fortin's inversion of the discrete divergence [69] will constantly be in the background, and we will just be repeating ideas that can be expressed in more abstract terms.)

It is very simple to see (no need of mixed variational formulation) that problem (2.14) has a unique solution $(\mathbf{q}, u) \in \mathbf{H}(\mathrm{div}, \Omega) \times H^1(\Omega)$, where

$$\mathbf{H}(\mathrm{div}, \Omega) := \{\mathbf{q} \in \mathbf{L}^2(\Omega) := L^2(\Omega)^d \,:\, \mathrm{div}\,\mathbf{q} \in L^2(\Omega)\}.$$

Discretization. For discretization let us consider a conforming partition $\mathscr{T}_h$ of Ω into triangles/tetrahedra, and the discrete spaces

$$\mathbf{V}_h := \prod_{K\in\mathscr{T}_h} \mathscr{RT}_k(K) = \{\mathbf{q}_h : \Omega \to \mathbb{R}^d \,:\, \mathbf{q}_h|_K \in \mathscr{RT}_k(K) \quad \forall K \in \mathscr{T}_h\}, \tag{2.15a}$$

$$W_h := \prod_{K\in\mathscr{T}_h} \mathscr{P}_k(K) = \{u_h : \Omega \to \mathbb{R} \,:\, u_h|_K \in \mathscr{P}_k(K) \quad \forall K \in \mathscr{T}_h\}, \tag{2.15b}$$

$$\mathbf{V}_h^{\mathrm{div}} := \mathbf{V}_h \cap \mathbf{H}(\mathrm{div}, \Omega). \tag{2.15c}$$

The RT approximation of (2.14) is a simple Galerkin scheme for a variational formulation of (2.14) obtained after integrating by parts in (2.14a), which naturally incorporates the BC (2.14c): we look for

$$(\mathbf{q}_h, u_h) \in \mathbf{V}_h^{\mathrm{div}} \times W_h, \tag{2.16a}$$

satisfying

$$(\kappa^{-1}\mathbf{q}_h, \mathbf{r})_\Omega - (u_h, \mathrm{div}\,\mathbf{r})_\Omega = -\langle g, \mathbf{r}\cdot\mathbf{n}\rangle_\Gamma \qquad \forall \mathbf{r} \in \mathbf{V}_h^{\mathrm{div}}, \tag{2.16b}$$

$$(\mathrm{div}\,\mathbf{q}_h, v)_\Omega = (f, v)_\Omega \qquad \forall v \in W_h. \tag{2.16c}$$

Existence and uniqueness of solution of (2.16) will follow from the arguments in the next section. Instead of using this Galerkin formulation, we will insert Lagrange multipliers to handle the continuity of the normal components of $\mathbf{q}_h$ across element interfaces: this leads to a formulation with three fields due to Arnold and Brezzi [1]. In Sect. 2.3, we will show how to eliminate the interior fields in order to build a discrete system that has lost the saddle point structure and only contains degrees of freedom on the faces.

2.2.1 The Arnold–Brezzi Formulation

Imposing continuity of the normal components. The key idea leading to the next equivalent presentation of Eqs. (2.16) is an observation about what condition functions in $\mathbf{V}_h$ must satisfy in order to belong to the space $\mathbf{V}_h^{\mathrm{div}}$. Let $K_1, K_2 \in \mathscr{T}_h$ meet in one face $\overline{K_1} \cap \overline{K_2} = \overline{e}$, with $e \in \mathscr{E}_h$. Given $\mathbf{q}_h \in \mathbf{V}_h$, it is easy to prove (based on Proposition 2.1(a)) that

$$\mathbf{q}_h \in \mathbf{V}_h^{\text{div}} \quad \Longrightarrow \quad \langle \mathbf{q}_h|_{K_1} \cdot \mathbf{n}_1 + \mathbf{q}_h|_{K_2} \cdot \mathbf{n}_2, \mu \rangle_e = 0 \quad \forall \mu \in \mathscr{P}_k(e). \tag{2.17}$$

Instead of writing the matching condition in the right-hand side of (2.17) for each interior $e \in \mathscr{E}_h$ looking for the elements on both sides of e, we can do as follows. For $\mathbf{q} : \Omega \to \mathbb{R}^d$ and $\mu : \cup\{\overline{e} \, : \, e \in \mathscr{E}_h\} \to \mathbb{R}$, we write

$$\langle \mathbf{q} \cdot \mathbf{n}, \mu \rangle_{\partial\mathscr{T}_h \setminus \Gamma} := \sum_{K \in \mathscr{T}_h} \langle \mathbf{q}|_K \cdot \mathbf{n}_K, \mu \rangle_{\partial K \setminus \Gamma}. \tag{2.18}$$

We then consider the space

$$M_h := \prod_{e \in \mathscr{E}_h} \mathscr{P}_k(e) = \{\mu : \cup\{\overline{e} \, : \, e \in \mathscr{E}_h\} \to \mathbb{R} \, : \, \mu|_e \in \mathscr{P}_k(e) \quad \forall e \in \mathscr{E}_h\}, \tag{2.19}$$

and finally group all conditions in (2.17) as

$$\langle \mathbf{q}_h \cdot \mathbf{n}, \mu \rangle_{\partial\mathscr{T}_h \setminus \Gamma} = 0 \qquad \forall \mu \in M_h. \tag{2.20}$$

This condition is then not only necessary but sufficient, that is, given $\mathbf{q}_h \in \mathbf{V}_h$, condition (2.20) is equivalent to the property $\mathbf{q}_h \in \mathbf{V}_h^{\text{div}}$. It is to be noticed that condition (2.20) does not use the entire space M_h but only the subspace

$$M_h^\circ := \{\mu \in M_h \, : \, \mu|_\Gamma = 0\}.$$

The remaining part is the space

$$M_h^\Gamma := \{\mu \in M_h \, : \, \mu|_e = 0 \quad \forall e \in \mathscr{E}_h^\circ\} \equiv \prod_{e \in \mathscr{E}_h \, e \subset \Gamma} \mathscr{P}_k(e),$$

where $\mathscr{E}_h^\circ$ is the set of interior faces.

Reaching the formulation. The side condition (2.18) will be compensated with the inclusion of a Lagrange multiplier, which will end up being an approximation of u on the skeleton of the triangulation (on the union of all faces of all the elements). Equation (2.16c) is naturally local, since the space W_h is a product space of polynomial spaces on the elements. Instead of using (2.16b), we will consider a similar equation based on each element. Note that we will not do any passage through the reference element in the remainder of this section, which will allow us to use the hat symbol to refer to a particular unknown of the discrete system. We then look for

$$(\mathbf{q}_h, u_h, \widehat{u}_h) \in \mathbf{V}_h \times W_h \times M_h, \tag{2.21a}$$

satisfying for all $K \in \mathscr{T}_h$

$$(\kappa^{-1}\mathbf{q}_h, \mathbf{r})_K - (u_h, \operatorname{div}\mathbf{r})_K + \langle \widehat{u}_h, \mathbf{r}\cdot\mathbf{n}\rangle_{\partial K} = 0 \qquad \forall \mathbf{r} \in \mathcal{RT}_k(K), \tag{2.21b}$$
$$(\operatorname{div}\mathbf{q}_h, w)_K = (f, w)_K \quad \forall w \in \mathcal{P}_k(K), \tag{2.21c}$$

as well as

$$\langle \mathbf{q}_h\cdot\mathbf{n}, \mu\rangle_{\partial\mathcal{T}_h\setminus\Gamma} = 0 \qquad \forall \mu \in M_h^\circ, \tag{2.21d}$$
$$\langle \widehat{u}_h, \mu\rangle_\Gamma = \langle g, \mu\rangle_\Gamma \qquad \forall \mu \in M_h^\Gamma. \tag{2.21e}$$

These equations can be written in global form using the following notation:

$$(u, v)_{\mathcal{T}_h} = \sum_{K\in\mathcal{T}_h} (u, v)_K, \qquad \langle \mathbf{q}\cdot\mathbf{n}, \mu\rangle_{\partial\mathcal{T}_h} = \sum_{K\in\mathcal{T}_h} \langle \mathbf{q}\cdot\mathbf{n}, \mu\rangle_{\partial K}.$$

(compare with (2.18) and adding the contributions of all the elements in the local Eqs. (2.21b) and (2.21c).) The $\mathcal{T}_h$-subscripted bracket will emphasize the fact that differential operators are applied element by element.

We look for

$$(\mathbf{q}_h, u_h, \widehat{u}_h) \in \mathbf{V}_h \times W_h \times M_h, \tag{2.22a}$$

satisfying

$$(\kappa^{-1}\mathbf{q}_h, \mathbf{r})_{\mathcal{T}_h} - (u_h, \operatorname{div}\mathbf{r})_{\mathcal{T}_h} + \langle \widehat{u}_h, \mathbf{r}\cdot\mathbf{n}\rangle_{\partial\mathcal{T}_h} = 0 \qquad \forall \mathbf{r} \in \mathbf{V}_h, \tag{2.22b}$$
$$(\operatorname{div}\mathbf{q}_h, w)_{\mathcal{T}_h} = (f, w)_{\mathcal{T}_h} \quad \forall w \in W_h, \tag{2.22c}$$
$$\langle \mathbf{q}_h\cdot\mathbf{n}, \mu\rangle_{\partial\mathcal{T}_h\setminus\Gamma} = 0 \quad \forall \mu \in M_h^\circ, \tag{2.22d}$$
$$\langle \widehat{u}_h, \mu\rangle_\Gamma = \langle g, \mu\rangle_\Gamma \quad \forall \mu \in M_h^\Gamma. \tag{2.22e}$$

Proposition 2.5 (Unique solvability)

(a) *Equations* (2.22) *are uniquely solvable.*
(b) *The solution of* (2.22) *solves* (2.16).
(c) *A solution of* (2.16) *can be added a field* $\widehat{u}_h \in M_h$ *to be a solution of* (2.22).

Proof Since $M_h \equiv M_h^\circ \oplus M_h^\Gamma$, existence of solution of (2.22) follows from uniqueness. Let then $(\mathbf{q}_h, u_h, \widehat{u}_h) \in \mathbf{V}_h \times W_h \times M_h$ be a solution of

$$(\kappa^{-1}\mathbf{q}_h, \mathbf{r})_{\mathcal{T}_h} - (u_h, \operatorname{div}\mathbf{r})_{\mathcal{T}_h} + \langle \widehat{u}_h, \mathbf{r}\cdot\mathbf{n}\rangle_{\partial\mathcal{T}_h} = 0 \quad \forall \mathbf{r} \in \mathbf{V}_h, \tag{2.23a}$$
$$(\operatorname{div}\mathbf{q}_h, w)_{\mathcal{T}_h} = 0 \quad \forall w \in W_h, \tag{2.23b}$$
$$\langle \mathbf{q}_h\cdot\mathbf{n}, \mu\rangle_{\partial\mathcal{T}_h\setminus\Gamma} = 0 \quad \forall \mu \in M_h^\circ, \tag{2.23c}$$
$$\langle \widehat{u}_h, \mu\rangle_\Gamma = 0 \quad \forall \mu \in M_h^\Gamma. \tag{2.23d}$$

Testing these equations with $(\mathbf{q}_h, u_h, -\widehat{u}_h, -\mathbf{q}_h\cdot\mathbf{n})$ and adding the results, we show that $(\kappa^{-1}\mathbf{q}_h, \mathbf{q}_h)_{\mathcal{T}_h} = 0$ and hence $\mathbf{q}_h = \mathbf{0}$. Let us now go back to (2.23a), which after integration by parts and localization on a single element yields for all $K \in \mathcal{T}_h$

$$-(\nabla u_h, \mathbf{r})_K + \langle u_h - \widehat{u}_h, \mathbf{r}\cdot\mathbf{n}\rangle_{\partial K} = 0 \qquad \forall \mathbf{r} \in \mathcal{RT}_k(K). \tag{2.24}$$

We now construct $\mathbf{p} \in \mathcal{RT}_k(K)$ satisfying

$$(\mathbf{p}, \mathbf{r})_K = 0 \qquad \forall \mathbf{r} \in \boldsymbol{\mathcal{P}}_{k-1}(K), \tag{2.25a}$$
$$\langle \mathbf{p}\cdot\mathbf{n}, \mu\rangle_{\partial K} = \langle u_h - \widehat{u}_h, \mu\rangle_{\partial K} \qquad \forall \mu \in \mathcal{R}_k(\partial K). \tag{2.25b}$$

(Note that these are the same equations that define the RT projection (2.6) and the local RT lifting of Sect. 2.1.3.) Using this function as the test function in (2.24), we prove that

$$0 = -(\nabla u_h, \mathbf{p})_K + \langle u_h - \widehat{u}_h, \mathbf{p}\cdot\mathbf{n}\rangle_{\partial K} = \langle u_h - \widehat{u}_h, u_h - \widehat{u}_h\rangle_{\partial K},$$

where we have used $\mu = u_h - \widehat{u}_h \in \mathcal{R}_k(\partial K)$ in (2.25b). Hence $u_h - \widehat{u}_h = 0$ on ∂K and (2.24) shows then (take $\mathbf{r} = \nabla u_h$) that $u_h \equiv c_K$ in K and $\widehat{u}_h = u_h \equiv c_K$ on ∂K. Since each interior face value of $\widehat{u}_h$ is reached from different elements, it is easy to see that we have proved that $u_h \equiv c$ and $\widehat{u}_h \equiv c$. However, Eq. (2.23d) implies that $\widehat{u}_h = 0$ on Γ, and the proof of uniqueness of solution of (2.22) is thus finished.

To prove (b), note first that (2.22d) implies that $\mathbf{q}_h \in \mathbf{V}_h^{\mathrm{div}}$. On the other hand, if we test Eq. (2.22b) with $\mathbf{r} \in \mathbf{V}_h^{\mathrm{div}} \subset \mathbf{V}_h$, it follows that

$$\begin{aligned} 0 &= (\kappa^{-1}\mathbf{q}_h, \mathbf{r})_{\mathcal{T}_h} - (u_h, \operatorname{div}\mathbf{r})_{\mathcal{T}_h} + \langle \widehat{u}_h, \mathbf{r}\cdot\mathbf{n}\rangle_{\partial\mathcal{T}_h} \\ &\ (\kappa^{-1}\mathbf{q}_h, \mathbf{r})_\Omega - (u_h, \operatorname{div}\mathbf{r})_\Omega + \langle \widehat{u}_h, \mathbf{r}\cdot\mathbf{n}\rangle_\Gamma && (\mathbf{r} \in \mathbf{V}_h^{\mathrm{div}}) \\ &\ (\kappa^{-1}\mathbf{q}_h, \mathbf{r})_\Omega - (u_h, \operatorname{div}\mathbf{r})_\Omega + \langle g, \mathbf{r}\cdot\mathbf{n}\rangle_\Gamma. && (\text{by (2.22e)}) \end{aligned}$$

This easily shows that any solution of (2.22) solves the traditional RT Eqs. (2.16).

Let now $(\mathbf{q}_h, u_h)$ solve (2.16). It is clear that Eqs. (2.22c) and (2.22d) are satisfied. We now look for $\widehat{u}_h \in M_h$ such that

$$\langle \widehat{u}_h, \mathbf{r}\cdot\mathbf{n}\rangle_{\partial\mathcal{T}_h} = -(\kappa^{-1}\mathbf{q}_h, \mathbf{r})_{\partial\mathcal{T}_h} + (u_h, \operatorname{div}\mathbf{r})_{\mathcal{T}_h} \qquad \forall \mathbf{r} \in \mathbf{V}_h. \tag{2.26}$$

The argument to show that (2.26) has a unique solution is simple. Uniqueness follows from the fact that if $\mu \in M_h$, there exists $\mathbf{r} \in \mathbf{V}_h$ such that $\langle \mu, \mathbf{r}\cdot\mathbf{n}\rangle_{\partial\mathcal{T}_h} = \langle \mu, \mu\rangle_{\partial\mathcal{T}_h}$ (this is done by using local liftings of the normal trace). To prove existence of solution, note that if $\mathbf{r} \in \mathbf{V}_h$ is such that $\langle \mu, \mathbf{r}\cdot\mathbf{n}\rangle_{\partial\mathcal{T}_h} = 0$ for all $\mu \in M_h$, then $\mathbf{r} \in \mathbf{V}_h^{\mathrm{div}}$ and $\mathbf{r}\cdot\mathbf{n} = 0$ on Γ, but in that case, by (2.16b), the right-hand side of (2.26) vanishes. This means that the right-hand side is orthogonal to the kernel of the transpose system. Then, by construction, (2.22b) is satisfied. Finally, if $\mu \in M_h^\Gamma$, we can easily find $\mathbf{r} \in \mathbf{V}_h^{\mathrm{div}}$ such that $\mathbf{r}\cdot\mathbf{n} = \mu$ on Γ. Then

$$
\begin{aligned}
\langle \widehat{u}_h, \mu\rangle_\Gamma &= \langle \widehat{u}_h, \mathbf{r}\cdot\mathbf{n}\rangle_\Gamma && \text{(by construction)}\\
&= \langle \widehat{u}_h, \mathbf{r}\cdot\mathbf{n}\rangle_{\partial\mathscr{T}_h} && (\mathbf{r}\in\mathbf{V}_h^{\mathrm{div}})\\
&= -(\kappa^{-1}\mathbf{q}_h, \mathbf{r})_{\mathscr{T}_h} + (u_h, \operatorname{div}\mathbf{r})_{\mathscr{T}_h} && \text{(by (2.26))}\\
&= \langle g, \mathbf{r}\cdot\mathbf{n}\rangle_\Gamma, && \text{(by (2.16b), since } \mathbf{r}\in\mathbf{V}_h^{\mathrm{div}})
\end{aligned}
$$

which is the missing equation in the decoupled formulation (2.22).

2.2.2 Energy Estimates

The error equations. Let us first recall the RT equations

$$
\begin{aligned}
(\kappa^{-1}\mathbf{q}_h, \mathbf{r})_{\mathscr{T}_h} - (u_h, \operatorname{div}\mathbf{r})_{\mathscr{T}_h} + \langle \widehat{u}_h, \mathbf{r}\cdot\mathbf{n}\rangle_{\partial\mathscr{T}_h} &= 0 && \forall \mathbf{r}\in\mathbf{V}_h, && \text{(2.27a)}\\
(\operatorname{div}\mathbf{q}_h, w)_{\mathscr{T}_h} &= (f, w)_{\mathscr{T}_h} && \forall w\in W_h, && \text{(2.27b)}\\
\langle \mathbf{q}_h\cdot\mathbf{n}, \mu\rangle_{\partial\mathscr{T}_h\setminus\Gamma} &= 0 && \forall \mu\in M_h^\circ, && \text{(2.27c)}\\
\langle \widehat{u}_h, \mu\rangle_\Gamma &= \langle g, \mu\rangle_\Gamma && \forall \mu\in M_h^\Gamma, && \text{(2.27d)}
\end{aligned}
$$

and let us note that these equations correspond to a consistent method:

$$
\begin{aligned}
(\kappa^{-1}\mathbf{q}, \mathbf{r})_{\mathscr{T}_h} - (u, \operatorname{div}\mathbf{r})_{\mathscr{T}_h} + \langle u, \mathbf{r}\cdot\mathbf{n}\rangle_{\partial\mathscr{T}_h} &= 0 && \forall \mathbf{r}\in\mathbf{V}_h, && \text{(2.28a)}\\
(\operatorname{div}\mathbf{q}, w)_{\mathscr{T}_h} &= (f, w)_{\mathscr{T}_h} && \forall w\in W_h, && \text{(2.28b)}\\
\langle \mathbf{q}\cdot\mathbf{n}, \mu\rangle_{\partial\mathscr{T}_h\setminus\Gamma} &= 0 && \forall \mu\in M_h^\circ, && \text{(2.28c)}\\
\langle u, \mu\rangle_\Gamma &= \langle g, \mu\rangle_\Gamma && \forall \mu\in M_h^\Gamma. && \text{(2.28d)}
\end{aligned}
$$

We then consider the projections $(\boldsymbol{\Pi}\mathbf{q}, \Pi u, \mathrm{P}u)\in\mathbf{V}_h\times W_h\times M_h$ defined by $\boldsymbol{\Pi}\mathbf{q}|_K := \boldsymbol{\Pi}^{\mathrm{RT}}\mathbf{q}$, $\Pi u|_K := \Pi_k u$ and

$$
\langle \mathrm{P}u, \mu\rangle_e = \langle u, \mu\rangle_e \qquad \forall \mu\in\mathscr{P}_k(e) \quad \forall e\in\mathscr{E}_h.
$$

Next, we substitute these projections in as many instances of (2.28) as possible:

$$
\begin{aligned}
(\kappa^{-1}\mathbf{q}, \mathbf{r})_{\mathscr{T}_h} - (\Pi u, \operatorname{div}\mathbf{r})_{\mathscr{T}_h} + \langle \mathrm{P}u, \mathbf{r}\cdot\mathbf{n}\rangle_{\partial\mathscr{T}_h} &= 0 && \forall \mathbf{r}\in\mathbf{V}_h, && \text{(2.29a)}\\
(\operatorname{div}\boldsymbol{\Pi}\mathbf{q}, w)_{\mathscr{T}_h} &= (f, w)_{\mathscr{T}_h} && \forall w\in W_h, && \text{(2.29b)}\\
\langle \boldsymbol{\Pi}\mathbf{q}\cdot\mathbf{n}, \mu\rangle_{\partial\mathscr{T}_h\setminus\Gamma} &= 0 && \forall \mu\in M_h^\circ, && \text{(2.29c)}\\
\langle \mathrm{P}u, \mu\rangle_\Gamma &= \langle g, \mu\rangle_\Gamma && \forall \mu\in M_h^\Gamma. && \text{(2.29d)}
\end{aligned}
$$

(Note that we have used the commutativity property (2.8) of the RT projection.) We then think in terms of the following quantities:

$$
\boldsymbol{\varepsilon}_h^q := \boldsymbol{\Pi}\mathbf{q} - \mathbf{q}_h \in \mathbf{V}_h, \quad \varepsilon_h^u := \Pi u - u_h \in W_h, \quad \widehat{\varepsilon}_h^u := \mathrm{P}u - \widehat{u}_h \in M_h. \qquad (2.30)
$$

Subtracting the discrete Eqs. (2.27) from (2.29), we get the **error equations**

$$
\begin{aligned}
(\kappa^{-1}\boldsymbol{\varepsilon}_h^q, \mathbf{r})_{\mathcal{T}_h} - (\varepsilon_h^u, \operatorname{div}\mathbf{r})_{\mathcal{T}_h} + \langle \widehat{\varepsilon}_h^u, \mathbf{r}\cdot\mathbf{n}\rangle_{\partial\mathcal{T}_h} &= (\kappa^{-1}(\boldsymbol{\Pi}\mathbf{q}-\mathbf{q}), \mathbf{r})_{\mathcal{T}_h}, && \text{(2.31a)}\\
(\operatorname{div}\boldsymbol{\varepsilon}_h^q, w)_{\mathcal{T}_h} &= 0, && \text{(2.31b)}\\
\langle \boldsymbol{\varepsilon}_h^q\cdot\mathbf{n}, \mu_1\rangle_{\partial\mathcal{T}_h\setminus\Gamma} &= 0, && \text{(2.31c)}\\
\langle \widehat{\varepsilon}_h^u, \mu_2\rangle_\Gamma &= 0, && \text{(2.31d)}
\end{aligned}
$$

for all $(\mathbf{r}, w, \mu_1, \mu_2) \in \mathbf{V}_h \times W_h \times M_h^\circ \times M_h^\Gamma$. Testing now Eqs. (2.31) with $(\boldsymbol{\varepsilon}_h^q, \varepsilon_h^u, -\widehat{\varepsilon}_h^u, -\boldsymbol{\varepsilon}_h^q\cdot\mathbf{n})$ and adding the results, we obtain the **energy identity**

$$
(\kappa^{-1}\boldsymbol{\varepsilon}_h^q, \boldsymbol{\varepsilon}_h^q)_{\mathcal{T}_h} = (\kappa^{-1}(\boldsymbol{\Pi}\mathbf{q}-\mathbf{q}), \boldsymbol{\varepsilon}_h^q)_{\mathcal{T}_h}. \tag{2.32}
$$

The Cauchy–Schwarz inequality w.r.t. the following norm

$$
\|\mathbf{q}\|_{\kappa^{-1}} = \|\kappa^{-1/2}\mathbf{q}\|_\Omega = (\kappa^{-1}\mathbf{q}, \mathbf{q})_\Omega^{1/2}
$$

provides our first convergence estimate

$$
\|\boldsymbol{\Pi}\mathbf{q} - \mathbf{q}_h\|_{\kappa^{-1}} = \|\boldsymbol{\varepsilon}_h^q\|_{\kappa^{-1}} \leqslant \|\boldsymbol{\Pi}\mathbf{q} - \mathbf{q}\|_{\kappa^{-1}}. \tag{2.33}
$$

On the decoupling in energy estimates. The estimate (2.33) is decoupled: convergence properties for $\mathbf{q}$ depend on approximation properties provided by the space $\mathbf{V}_h$, but not on those of the space W_h. This is not the case for equations of the form

$$
\operatorname{div}\mathbf{q} + c\,u = f, \qquad \text{where } c \geqslant 0.
$$

We will study this effect in Sect. 2.3.5.

A flux estimate. For any $\mathbf{p} \in \mathcal{P}_{k+1}(K)$, we have

$$
\begin{aligned}
h_K\langle \mathbf{p}\cdot\mathbf{n}, \mathbf{p}\cdot\mathbf{n}\rangle_{\partial K} &= h_K\langle \widetilde{\mathbf{p}\cdot\mathbf{n}}, \widetilde{\mathbf{p}\cdot\mathbf{n}}\rangle_{\partial\widehat{K}} && \text{(by (1.1c))}\\
&= h_K\langle \widetilde{\mathbf{p}\cdot\mathbf{n}}, |a_K|^{-1}\widetilde{\mathbf{p}\cdot\mathbf{n}}\rangle_{\partial\widehat{K}} \\
&= h_K\langle \widehat{\mathbf{p}}\cdot\widehat{\mathbf{n}}, |a_K|^{-1}\widehat{\mathbf{p}}\cdot\widehat{\mathbf{n}}\rangle_{\partial\widehat{K}} && \text{(by (1.3c))}\\
&\lesssim h_K^{2-d}\|\widehat{\mathbf{p}}\cdot\widehat{\mathbf{n}}\|_{\partial\widehat{K}}^2 && \text{(by (1.6))}\\
&\lesssim h_K^{2-d}\|\widehat{\mathbf{p}}\|_{\widehat{K}}^2 && \text{(finite dimension)}\\
&\approx \|\mathbf{p}\|_K^2. && \text{(by (1.7))}
\end{aligned}
$$

If we then use the norm

$$
\|\mu\|_h := \left(\sum_{K\in\mathcal{T}_h} h_K\|\mu\|_{\partial K}^2\right)^{1/2} \approx \left(\sum_{e\in\mathcal{E}_h} h_e\|\mu\|_e^2\right)^{1/2},
$$

we have a bound

$$\|\boldsymbol{\varepsilon}_h^q \cdot \mathbf{n}\|_h \lesssim \|\boldsymbol{\varepsilon}_h^q\|_\Omega. \tag{2.34}$$

2.2.3 More Convergence Estimates

Lifting ε_h^u. The key to an error analysis of u_h is a lifting to ε_h^u to be the divergence of a continuous vector field. Suppose there exists an operator $\mathbf{L} : L^2(\Omega) \to \mathbf{H}^1(\Omega)$ such that

$$\operatorname{div} \mathbf{L} v = v, \qquad \|\mathbf{L}\, v\|_{1,\Omega} \lesssim \|v\|_\Omega.$$

Note that the existence of $\mathbf{L}$ is obviously true if we assume the regularity hypothesis (2.42). Let $\boldsymbol{\xi} := \mathbf{L}\varepsilon_h^u$, so that

$$\operatorname{div} \boldsymbol{\xi} = \varepsilon_h^u, \qquad \|\boldsymbol{\xi}\|_{1,\Omega} \lesssim \|\varepsilon_h^u\|_\Omega. \tag{2.35}$$

The error analysis is then based on using $\boldsymbol{\Pi}\boldsymbol{\xi}$ as test function in (2.29a) and (2.27a) and subtracting the result:

$$(\kappa^{-1}(\mathbf{q}_h - \mathbf{q}), \boldsymbol{\Pi}\boldsymbol{\xi})_{\mathscr{T}_h} - (\varepsilon_h^u, \operatorname{div} \boldsymbol{\Pi}\boldsymbol{\xi})_{\mathscr{T}_h} + \langle \widehat{\varepsilon}_h^u, (\boldsymbol{\Pi}\boldsymbol{\xi}) \cdot \mathbf{n}\rangle_{\partial\mathscr{T}_h} = 0. \tag{2.36}$$

We then go ahead and study what is in (2.36). First of all

$$\begin{aligned}
\langle \widehat{\varepsilon}_h^u, (\boldsymbol{\Pi}\boldsymbol{\xi}) \cdot \mathbf{n}\rangle_{\partial\mathscr{T}_h} &= \langle \widehat{\varepsilon}_h^u, \boldsymbol{\xi} \cdot \mathbf{n}\rangle_{\partial\mathscr{T}_h} && \text{(definition of the RT projection)}\\
&= \langle \widehat{\varepsilon}_h^u, \boldsymbol{\xi} \cdot \mathbf{n}\rangle_{\partial\mathscr{T}_h \setminus \Gamma} && (\widehat{\varepsilon}_h^u = 0 \text{ on } \Gamma \text{ by (2.31d)})\\
&= 0. && (\boldsymbol{\xi} \cdot \mathbf{n} \text{ changes sign on internal faces})
\end{aligned}$$

Therefore

$$\begin{aligned}
\|\varepsilon_h^u\|_\Omega^2 &= (\varepsilon_h^u, \operatorname{div} \boldsymbol{\xi})_{\mathscr{T}_h} && \text{(by (2.35))}\\
&= (\varepsilon_h^u, \Pi \operatorname{div} \boldsymbol{\xi})_{\mathscr{T}_h} && (\varepsilon_h^u \in W_h)\\
&= (\varepsilon_h^u, \operatorname{div} \boldsymbol{\Pi}\boldsymbol{\xi})_{\mathscr{T}_h} && \text{(commutativity property (2.8))}\\
&= (\kappa^{-1}(\mathbf{q}_h - \mathbf{q}), \boldsymbol{\Pi}\boldsymbol{\xi})_{\mathscr{T}_h} && \text{(by the error Eq. (2.36))}\\
&\leqslant \|\mathbf{q} - \mathbf{q}_h\|_{\kappa^{-1}} \|\kappa^{-1/2}\|_{L^\infty} \|\boldsymbol{\Pi}\boldsymbol{\xi}\|_\Omega\\
&\lesssim \|\kappa^{-1/2}\|_{L^\infty} \|\mathbf{q} - \mathbf{q}_h\|_{\kappa^{-1}} \|\boldsymbol{\xi}\|_{1,\Omega} && \text{(by Proposition 2.3)}\\
&\lesssim \|\mathbf{q} - \mathbf{q}_h\|_{\kappa^{-1}} \|\varepsilon_h^u\|_\Omega. && \text{(by (2.35))}
\end{aligned}$$

This takes us to our second error estimate

$$\|\Pi u - u_h\|_\Omega = \|\varepsilon_h^u\|_\Omega \lesssim \|\mathbf{q} - \mathbf{q}_h\|_{\kappa^{-1}}. \tag{2.37}$$

Lifting $\widehat{\varepsilon}_h^u$ **locally**. The error analysis for $\widehat{u}_h$ is carried out on an element-by-element basis using the lifting operator of Proposition 2.4. Let then $\mathbf{r} := \mathbf{L}^{\mathrm{RT}} \widehat{\varepsilon}_h^u|_{\partial K} \in \mathscr{R}\mathscr{T}_k(K)$, so that

$$\mathbf{r}\cdot\mathbf{n}=\widehat{\varepsilon}_h^u, \qquad \|\mathbf{r}\|_K \lesssim h_K^{1/2}\|\widehat{\varepsilon}_h^u\|_{\partial K}. \tag{2.38}$$

Also, using a scaling argument (and the fact that $\mathbf{r}$ is in a polynomial space), we show that

$$h_K|\mathbf{r}|_{1,K} \lesssim \|\mathbf{r}\|_K \lesssim h_K^{1/2}\|\widehat{\varepsilon}_h^u\|_{\partial K}. \tag{2.39}$$

Therefore

$$\begin{aligned}
\|\widehat{\varepsilon}_h^u\|_{\partial K}^2 &= \langle\widehat{\varepsilon}_h^u, \mathbf{r}\cdot\mathbf{n}\rangle_{\partial K} && \text{(by (2.38))}\\
&= (\varepsilon_h^u, \operatorname{div}\mathbf{r})_K - (\kappa^{-1}(\mathbf{q}-\mathbf{q}_h), \mathbf{r})_K && \text{(by error eqn (2.31a))}\\
&\lesssim h_K^{-1/2}\Big(\|\varepsilon_h^u\|_K + h_K\|\kappa^{-1/2}(\mathbf{q}-\mathbf{q}_h)\|_K\Big)\|\widehat{\varepsilon}_h^u\|_{\partial K}. && \text{(by (2.38) \& (2.39))}
\end{aligned}$$

A complete estimate can now be proved by adding the previous inequalities over all triangles. We have then essentially proved that

$$\|\mathrm{P}u - \widehat{u}_h\|_h = \|\widehat{\varepsilon}_h^u\|_h \lesssim \|\varepsilon_h^u\|_\Omega + h\|\mathbf{q}-\mathbf{q}_h\|_{\kappa^{-1}}. \tag{2.40}$$

2.2.4 *Superconvergence Estimates by Duality*

Another inverse of the divergence. A superconvergence analysis for u_h can be carried out by using a more demanding form of writing $\operatorname{div}\boldsymbol{\xi}=\varepsilon_h^u$ than the one used in (2.35). In particular, we will also be using the second of the error Eqs. (2.31) in the arguments that follow. We start by considering a dual problem

$$\kappa^{-1}\boldsymbol{\xi} - \nabla\theta = \mathbf{0} \quad \text{in } \Omega, \tag{2.41a}$$
$$\operatorname{div}\boldsymbol{\xi} = \varepsilon_h^u \quad \text{in } \Omega, \tag{2.41b}$$
$$\theta = 0 \quad \text{on } \Gamma. \tag{2.41c}$$

We assume the following **regularity hypothesis**: there exists $C_{\mathrm{reg}} > 0$ such that

$$\|\boldsymbol{\xi}\|_{1,\Omega} + \|\theta\|_{2,\Omega} \leqslant C_{\mathrm{reg}}\|\varepsilon_h^u\|_\Omega. \tag{2.42}$$

This estimate holds for convex domains with smooth diffusion coefficient κ.

The duality estimate. The beginning of the argument can be copied verbatim from what we did in Sect. 2.2.3:

$$\begin{aligned}
\|\varepsilon_h^u\|_\Omega^2 &= (\kappa^{-1}(\mathbf{q}_h-\mathbf{q}), \boldsymbol{\Pi}\boldsymbol{\xi})_{\mathscr{T}_h}\\
&= (\kappa^{-1}(\mathbf{q}_h-\mathbf{q}), \boldsymbol{\Pi}\boldsymbol{\xi}-\boldsymbol{\xi})_{\mathscr{T}_h} + (\mathbf{q}_h-\mathbf{q}, \kappa^{-1}\boldsymbol{\xi})_{\mathscr{T}_h}\\
&= (\kappa^{-1}(\mathbf{q}_h-\mathbf{q}), \boldsymbol{\Pi}\boldsymbol{\xi}-\boldsymbol{\xi})_{\mathscr{T}_h} + (\mathbf{q}_h-\mathbf{q}, \nabla\theta)_{\mathscr{T}_h}. && \text{(by (2.41a))}
\end{aligned}$$

The next part of the argument consists of working on the rightmost term in the previous inequality. This yields

$$
\begin{aligned}
(\mathbf{q}_h - \mathbf{q}, \nabla\theta)_{\mathcal{T}_h} &= -(\mathrm{div}\,(\mathbf{q}_h - \mathbf{q}), \theta)_{\mathcal{T}_h} \\
&\quad + \langle(\mathbf{q}_h - \mathbf{q})\cdot\mathbf{n}, \theta\rangle_{\partial\mathcal{T}_h\setminus\Gamma} && (\theta = 0 \text{ on } \Gamma) \\
&= (\mathrm{div}\,(\mathbf{q} - \mathbf{q}_h), \theta)_{\mathcal{T}_h} && (\text{single-valued on } \partial\mathcal{T}_h) \\
&= (f - \Pi f, \theta)_\Omega && (\mathrm{div}\mathbf{q}_h = \Pi f \text{ is (2.27}b\text{))} \\
&= (f - \Pi f, \theta - \Pi\theta)_\Omega.
\end{aligned}
$$

We end up by putting everything together and using estimates of the projections

$$
\begin{aligned}
\|\varepsilon_h^u\|_\Omega^2 &= (\kappa^{-1}(\mathbf{q}_h - \mathbf{q}), \boldsymbol{\Pi}\boldsymbol{\xi} - \boldsymbol{\xi})_{\mathcal{T}_h} + (\mathbf{q}_h - \mathbf{q}, \nabla\theta)_{\mathcal{T}_h} \\
&= (\kappa^{-1}(\mathbf{q}_h - \mathbf{q}), \boldsymbol{\Pi}\boldsymbol{\xi} - \boldsymbol{\xi})_{\mathcal{T}_h} + (f - \Pi f, \theta - \Pi\theta)_\Omega \\
&\lesssim h\|\mathbf{q}_h - \mathbf{q}\|_{\kappa^{-1}}|\boldsymbol{\xi}|_{1,\Omega} + h\|f - \Pi f\|_\Omega|\theta|_{1,\Omega} \\
&\lesssim h\,(\|\mathbf{q}_h - \mathbf{q}\|_{\kappa^{-1}} + \|f - \Pi f\|_\Omega)\|\varepsilon_h^u\|_\Omega. && (\text{reg. hypothesis (2.42)})
\end{aligned}
$$

(The argument uses that $\|\mathbf{p} - \boldsymbol{\Pi}^{\mathrm{RT}}\mathbf{p}\|_K \lesssim h_K|\mathbf{p}|_{1,K}$. This can easily be proved using the same arguments as in Proposition 2.3(b).) We have thus proved that, under the regularity hypothesis (2.42),

$$
\|\varepsilon_h^u\|_\Omega \lesssim h\left(\|\mathbf{q}_h - \mathbf{q}\|_{\kappa^{-1}} + \|f - \Pi f\|_\Omega\right). \tag{2.43}
$$

This bound can then be used in the right-hand side of (2.40) to show that

$$
\|\widehat{\varepsilon}_h^u\|_h \lesssim h\left(\|\mathbf{q}_h - \mathbf{q}\|_{\kappa^{-1}} + \|f - \Pi f\|_\Omega\right).
$$

2.2.5 Summary of Estimates

Approximation properties. Let us start by recalling that

$$
\|\mathbf{q} - \boldsymbol{\Pi}\mathbf{q}\|_\Omega \lesssim h^{k+1}|\mathbf{q}|_{k+1,\Omega} \qquad \text{and} \qquad \|u - \Pi u\|_\Omega \lesssim h^{k+1}|u|_{k+1,\Omega}.
$$

Also,

$$
\begin{aligned}
h_K^{\frac{1}{2}}\|u - \mathrm{P}u\|_{\partial K} &\lesssim h_K^{\frac{d}{2}}\|\widehat{u} - \widehat{\mathrm{P}u}\|_{\partial\widehat{K}} && (\text{change of variables (1.7)}) \\
&= h_K^{\frac{d}{2}}\|\widehat{u} - \widehat{\mathrm{P}}\widehat{u}\|_{\partial\widehat{K}} && (\text{easy argument}) \\
&\leqslant h_K^{\frac{d}{2}}\|\widehat{u} - \widehat{\Pi}_k\widehat{u}\|_{\partial\widehat{K}} && (\widehat{\mathrm{P}} \text{ gives the best approx}) \\
&\lesssim h_K^{\frac{d}{2}}\|\widehat{u} - \widehat{\Pi}_k\widehat{u}\|_{1,\widehat{K}} && (\text{trace theorem}) \\
&\lesssim h_K^{\frac{d}{2}}|\widehat{u}|_{k+1,\widehat{K}} && (\text{compactness}) \\
&\lesssim h_K^{k+1}|u|_{k+1,K}, && (\text{change of variables (1.9)})
\end{aligned}
$$

which can be collected in the estimate

$$\|u - \mathrm{P}u\|_h \lesssim h^{k+1}|u|_{k+1,\Omega}.$$

It is also easy to see that

$$\|\mathbf{q}\cdot\mathbf{n} - \boldsymbol{\Pi}\mathbf{q}\cdot\mathbf{n}\|_h \lesssim h^{k+1}|\mathbf{q}|_{k+1,\Omega}.$$

This is done element by element, face by face, using the fact that $\boldsymbol{\Pi}\mathbf{q}\cdot\mathbf{n}|_{\partial K}$ is the best approximation of $\mathbf{q}\cdot\mathbf{n}$ on $\mathscr{R}_k(\partial K)$ and, therefore, we can use the previous estimate applied to $u = \mathbf{q}\cdot\mathbf{n}_e$ for every $e \in \mathscr{E}(K)$.

Optimal convergence. Assuming that everything is going the best way it can (solutions are smooth, the regularity hypotheses holds), we can summarize the convergence results in the following table:

$$\begin{aligned}
\|\mathbf{q}-\mathbf{q}_h\|_\Omega &\lesssim h^{k+1}, & \|\boldsymbol{\Pi}\mathbf{q}-\mathbf{q}_h\|_\Omega &\lesssim h^{k+1}, && \text{(by (2.33))}\\
\|u-u_h\|_\Omega &\lesssim h^{k+1}, & \|\Pi u-u_h\|_\Omega &\lesssim h^{k+2}, && \text{(by (2.43))}\\
\|u-\widehat{u}_h\|_h &\lesssim h^{k+1}, & \|\mathrm{P}u-\widehat{u}_h\|_h &\lesssim h^{k+2}, && \text{(by (2.40) \& (2.43))}\\
\|\mathbf{q}\cdot\mathbf{n}-\mathbf{q}_h\cdot\mathbf{n}\|_h &\lesssim h^{k+1}, & \|\boldsymbol{\Pi}\mathbf{q}\cdot\mathbf{n}-\mathbf{q}_h\cdot\mathbf{n}\|_h &\lesssim h^{k+1}. && \text{(by (2.34) \& (2.33))}
\end{aligned}$$

2.3 Additional Topics

The following section explains some topics that are related to the RT method, or more specifically to the Arnold–Brezzi formulation of the RT method. These are general ideas that will apply with minimal changes to the other two methods (BDM and HDG) that we will introduce in these notes. For reasons of notation, we will write the diffusion problem as

$$\kappa^{-1}\mathbf{q} + \nabla u = \mathbf{0} \quad \text{in } \Omega, \tag{2.44a}$$

$$\operatorname{div}\mathbf{q} = f \quad \text{in } \Omega, \tag{2.44b}$$

$$u = g \quad \text{on } \Gamma. \tag{2.44c}$$

The term hybridization makes reference to the not-that-popular hybrid methods, where the variational formulation is taken directly on the interfaces of the elements. (Yes, some of them can be understood as domain decomposition methods, and yes, the ultra-weak variational formulation UWVF is also related.) For more about hybrid methods—that we will not touch here—the reader is referred to Brezzi and Fortin's book [10].

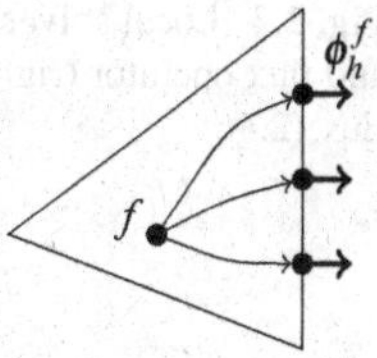

Fig. 2.1 Flux ϕ_h^f due to sources f. See Eqs. (2.45)

2.3.1 Hybridization

What is hybridization? The goal of hybridization is the reduction of the system (2.22) to a linear system where only $\widehat{u}_h$ shows up [41]. The remaining two variables will be reconstructed after solving for $\widehat{u}_h$, in an element-by-element fashion. This is easy to realize due to the fact that Eqs. (2.22b) and (2.22c) are local or, in other words, the spaces $\mathbf{V}_h$ and W_h are completely discontinuous. For some forthcoming arguments, it'll be practical to deal with the space

$$B_h := \prod_{K \in \mathscr{T}_h} \mathscr{R}_k(\partial K),$$

and to note that M_h is the subset of B_h of functions that are single-valued.

Flux due to sources. Given $f : \Omega \to \mathbb{R}$, we look for

$$(\mathbf{q}_h^f, u_h^f) \in \mathbf{V}_h \times W_h, \tag{2.45a}$$

satisfying

$$(\kappa^{-1}\mathbf{q}_h^f, \mathbf{r})_{\mathscr{T}_h} - (u_h^f, \operatorname{div} \mathbf{r})_{\mathscr{T}_h} = 0 \qquad \forall \mathbf{r} \in \mathbf{V}_h, \tag{2.45b}$$

$$(\operatorname{div} \mathbf{q}_h^f, w)_{\mathscr{T}_h} = (f, w)_{\mathscr{T}_h} \qquad \forall w \in W_h. \tag{2.45c}$$

(Existence and uniqueness of solution of (2.45) is straightforward to prove.) We then define

$$\phi_h^f := -\mathbf{q}_h^f \cdot \mathbf{n} \in B_h. \tag{2.45d}$$

Local solvers and flux operators. Consider now the operator

$$M_h \ni \widehat{u}_h \quad \longmapsto \quad (\mathrm{L}^q(\widehat{u}_h), \mathrm{L}^u(\widehat{u}_h)) = (\mathbf{q}_h, u_h) \in \mathbf{V}_h \times W_h, \tag{2.46a}$$

where

$$(\kappa^{-1}\mathbf{q}_h, \mathbf{r})_{\mathscr{T}_h} - (u_h, \operatorname{div} \mathbf{r})_{\mathscr{T}_h} + \langle \widehat{u}_h, \mathbf{r} \cdot \mathbf{n} \rangle_{\partial \mathscr{T}_h} = 0 \qquad \forall \mathbf{r} \in \mathbf{V}_h, \tag{2.46b}$$

$$(\operatorname{div} \mathbf{q}_h, w)_{\mathscr{T}_h} = 0 \qquad \forall w \in W_h. \tag{2.46c}$$

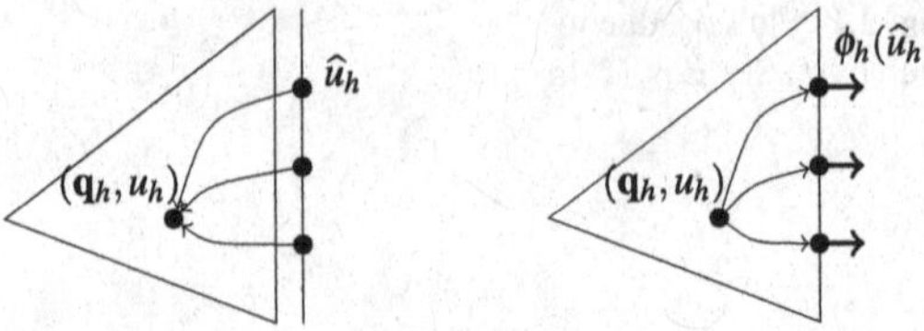

Fig. 2.2 Local solver (left) and flux operator (right). See Eqs. (2.46)

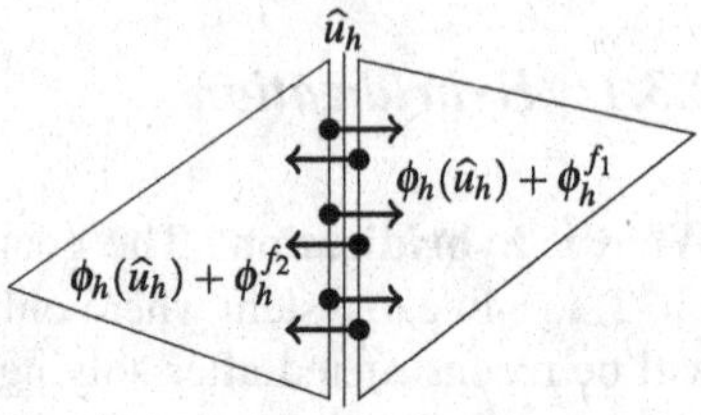

Fig. 2.3 The flux coming from left to right $\phi_h(\widehat{u}_h) + \phi_h^{f_1}$ cancels the flux from right to left $\phi_h(\widehat{u}_h) + \phi_h^{f_2}$. This is nothing but Eq. (2.47b)

We then consider the flux operator $\phi_h : M_h \to B_h$ given by

$$\phi_h(\widehat{u}_h) := -\mathbf{q}_h \cdot \mathbf{n}. \tag{2.46d}$$

Note that Eqs. (2.46b)–(2.46c) are uniquely solvable and can be solved element by element.

The hybridized system. We look for

$$\widehat{u}_h \in M_h, \tag{2.47a}$$

satisfying

$$\langle \phi_h(\widehat{u}_h) + \phi_h^f, \mu \rangle_{\partial \mathscr{T}_h \setminus \Gamma} = 0 \qquad \forall \mu \in M_h^\circ, \tag{2.47b}$$

$$\langle \widehat{u}_h, \mu \rangle_\Gamma = \langle g, \mu \rangle_\Gamma \qquad \forall \mu \in M_h^\Gamma. \tag{2.47c}$$

We then define

$$\mathbf{q}_h = \mathrm{L}^q(\widehat{u}_h) + \mathbf{q}_h^f, \qquad u_h = \mathrm{L}^u(\widehat{u}_h) + u_h^f. \tag{2.47d}$$

Note that if we subtract

$$\widehat{u}_h^g \in M_h^\Gamma \quad \text{satisfying} \quad \langle \widehat{u}_h^g, \mu \rangle_\Gamma = \langle g, \mu \rangle_\Gamma \quad \forall \mu \in M_h^\Gamma,$$

then the hybridized system can be written as

$$\widehat{u}_h^\circ \in M_h^\circ \ \text{ s.t. } \ \langle \phi_h(\widehat{u}_h^\circ), \mu \rangle_{\partial \mathscr{T}_h \setminus \Gamma} = -\langle \phi_h^f + \phi_h(\widehat{u}_h^g), \mu \rangle_{\partial \mathscr{T}_h \setminus \Gamma} \qquad \forall \mu \in M_h^\circ.$$

The hybridized bilinear form. We next focus on the bilinear form

$$M_h^\circ \times M_h^\circ \ni (\lambda, \mu) \quad \longmapsto \quad \langle \phi_h(\lambda), \mu \rangle_{\partial \mathscr{T}_h \setminus \Gamma}. \tag{2.48}$$

Let then $(\mathbf{q}_h, u_h) = (\mathrm{L}^q(\lambda), \mathrm{L}^u(\lambda))$ and $(\mathbf{v}_h, v_h) = (\mathrm{L}^q(\mu), \mathrm{L}^u(\mu))$. Note that

$$\begin{aligned} (\kappa^{-1}\mathbf{v}_h, \mathbf{r})_{\mathscr{T}_h} - (v_h, \operatorname{div}\mathbf{r})_{\mathscr{T}_h} + \langle \mu, \mathbf{r}\cdot\mathbf{n}\rangle_{\partial\mathscr{T}_h} &= 0 \quad \forall \mathbf{r} \in \mathbf{V}_h,\\ (\operatorname{div}\mathbf{q}_h, w)_{\mathscr{T}_h} &= 0 \quad \forall w \in W_h, \end{aligned}$$

and therefore

$$\begin{aligned} (\kappa^{-1}\mathbf{v}_h, \mathbf{q}_h)_{\mathscr{T}_h} - (v_h, \operatorname{div}\mathbf{q}_h)_{\mathscr{T}_h} &= -\langle \mu, \mathbf{q}_h\cdot\mathbf{n}\rangle_{\partial\mathscr{T}_h},\\ (\operatorname{div}\mathbf{q}_h, v_h)_{\mathscr{T}_h} &= 0, \end{aligned}$$

which implies that

$$\begin{aligned} \langle \phi_h(\lambda), \mu\rangle_{\partial\mathscr{T}_h\setminus\Gamma} &= \langle \phi_h(\lambda), \mu\rangle_{\partial\mathscr{T}_h} && (\mu \in M_h^\circ)\\ &= -\langle \mathbf{q}_h\cdot\mathbf{n}, \mu\rangle_{\partial\mathscr{T}_h} && (\text{definition of } \phi_h)\\ &= (\kappa^{-1}\mathbf{v}_h, \mathbf{q}_h)_{\mathscr{T}_h}. \end{aligned}$$

It is clear from this expression that the bilinear form is **symmetric** and positive semidefinite. On the other hand, if $\lambda \in M_h^\circ$ and $\langle \phi_h(\lambda), \lambda\rangle_{\partial\mathscr{T}_h\setminus\Gamma} = 0$, it is a simple exercise to observe that $(L^q(\lambda), L^u(\lambda), \lambda)$ is a solution of the discrete Eqs. (2.22) with zero right-hand side and therefore has to vanish. This proves the **positive definiteness** of the bilinear form (2.48).

2.3.2 A Discrete Dirichlet Form

Toward a primal form. The goal of this section is the proof that the system (2.22) can be written in the variable u_h only. This is not useful from the practical point of view but helps in arguments related to the RT discretization of evolutionary partial differential equations.

Lifting of Dirichlet conditions. Given $g : \Gamma \to \mathbb{R}$, we consider the pair

$$(\mathbf{q}_h^g, \widehat{u}_h^g) \in \mathbf{V}_h \times M_h, \tag{2.49a}$$

satisfying

$$\begin{aligned} (\kappa^{-1}\mathbf{q}_h^g, \mathbf{r})_{\mathscr{T}_h} + \langle \widehat{u}_h^g, \mathbf{r}\cdot\mathbf{n}\rangle_{\partial\mathscr{T}_h} &= 0 && \forall \mathbf{r} \in \mathbf{V}_h, && (2.49b)\\ \langle \mathbf{q}_h^g\cdot\mathbf{n}, \mu\rangle_{\partial\mathscr{T}_h\setminus\Gamma} &= 0 && \forall \mu \in M_h^\circ, && (2.49c)\\ \langle \widehat{u}_h^g, \mu\rangle_\Gamma &= \langle g, \mu\rangle_\Gamma && \forall \mu \in M_h^\Gamma. && (2.49d) \end{aligned}$$

(Existence and uniqueness of solutions to this problem is an easy exercise.) We then define

$$W_h \ni w \quad \longmapsto \quad \ell(g, w) := (\operatorname{div}\mathbf{q}_h^g, w)_{\mathscr{T}_h}. \tag{2.49e}$$

The RT gradient. We now consider the map

$$W_h \ni u_h \quad \longmapsto \quad (\mathscr{G}_h^q u_h, \mathscr{G}_h^{\widehat{u}} u_h) = (\mathbf{q}_h, \widehat{u}_h) \in \mathbf{V}_h \times M_h, \tag{2.50a}$$

where

$$(\kappa^{-1}\mathbf{q}_h, \mathbf{r})_{\mathscr{T}_h} - (u_h, \operatorname{div} \mathbf{r})_{\mathscr{T}_h} + \langle \widehat{u}_h, \mathbf{r} \cdot \mathbf{n} \rangle_{\partial \mathscr{T}_h} = 0 \qquad \forall \mathbf{r} \in \mathbf{V}_h, \tag{2.50b}$$

$$\langle \mathbf{q}_h \cdot \mathbf{n}, \mu \rangle_{\partial \mathscr{T}_h \setminus \Gamma} = 0 \qquad \forall \mu \in M_h^\circ, \tag{2.50c}$$

$$\langle \widehat{u}_h, \mu \rangle_\Gamma = 0 \qquad \forall \mu \in M_h^\Gamma. \tag{2.50d}$$

We can thus think of the bilinear form (the discrete Dirichlet form)

$$(u_h, v_h) \ni W_h \times W_h \quad \longmapsto \quad D_h(u_h, v_h) := (\operatorname{div} \mathbf{q}_h, v_h)_{\mathscr{T}_h} = (\operatorname{div} \mathscr{G}_h^q u_h, v_h)_{\mathscr{T}_h}. \tag{2.50e}$$

Note that $\mathscr{G}_h^q$ is a minus gradient operator, instead of a gradient operator.

The primal form. Given $f : \Omega \to \mathbb{R}$ and $g : \Gamma \to \mathbb{R}$, we look for

$$u_h \in W_h \quad \text{satisfying} \quad D_h(u_h, w) = (f, w)_{\mathscr{T}_h} - \ell(g, w) \quad \forall w \in W_h.$$

This implies that

$$\mathbf{q}_h = \mathbf{q}_h^g + \mathscr{G}_h^q u_h, \qquad \widehat{u}_h = \widehat{u}_h^g + \mathscr{G}_h^{\widehat{u}} u_h,$$

and u_h constitute the solution of (2.22). It is not difficult to figure out that the primal form is just the Schur complement form of the traditional RT formulation (2.16).

Properties of the Dirichlet form. Given $(u_h, v_h) \in W_h \times W_h$, we consider $(\mathbf{q}_h, \widehat{u}_h) = (\mathscr{G}_h^q u_h, \mathscr{G}_h^{\widehat{u}} u_h)$ and $(\mathbf{v}_h, \widehat{v}_h) = (\mathscr{G}_h^q v_h, \mathscr{G}_h^{\widehat{u}} v_h)$. Note that

$$(\kappa^{-1}\mathbf{v}_h, \mathbf{r})_{\mathscr{T}_h} - (v_h, \operatorname{div} \mathbf{r})_{\mathscr{T}_h} + \langle \widehat{v}_h, \mathbf{r} \cdot \mathbf{n} \rangle_{\partial \mathscr{T}_h \setminus \Gamma} = 0 \qquad \forall \mathbf{r} \in \mathbf{V}_h,$$

$$\langle \mathbf{q}_h \cdot \mathbf{n}, \mu \rangle_{\partial \mathscr{T}_h \setminus \Gamma} = 0 \qquad \forall \mu \in M_h^\circ,$$

and therefore

$$(\kappa^{-1}\mathbf{v}_h, \mathbf{q}_h)_{\mathscr{T}_h} - (v_h, \operatorname{div} \mathbf{q}_h)_{\mathscr{T}_h} + \langle \widehat{v}_h, \mathbf{q}_h \cdot \mathbf{n} \rangle_{\partial \mathscr{T}_h \setminus \Gamma} = 0,$$

$$\langle \mathbf{q}_h \cdot \mathbf{n}, \widehat{v}_h \rangle_{\partial \mathscr{T}_h \setminus \Gamma} = 0.$$

We have

$$D_h(u_h, v_h) = (\operatorname{div} \mathbf{q}_h, v_h)_{\mathscr{T}_h} = (\kappa^{-1}\mathbf{v}_h, \mathbf{q}_h)_{\mathscr{T}_h},$$

which proves that the discrete Dirichlet form is **symmetric** and positive semidefinite. Now, if $D(u_h, u_h) = 0$, it is easy to see how $(\mathscr{G}_h^q u_h, u_h, \mathscr{G}_h^{\widehat{u}} u_h)$ is a solution of (2.22) with zero right-hand side, and therefore it has to vanish, which proves that the discrete Dirichlet form is **positive definite**.

2.3.3 Stenberg Postprocessing

The local postprocessing step. Assume that we have solved the RT Eqs. (2.22). We look for

$$u_h^\star \in \prod_{K\in\mathcal{T}_h} \mathcal{P}_{k+1}(K), \tag{2.51a}$$

satisfying for all $K \in \mathcal{T}_h$

$$\begin{aligned}(\kappa\nabla u_h^\star, \nabla v)_K &= (f, v)_K - \langle \mathbf{q}_h\cdot\mathbf{n}, v\rangle_{\partial K} \qquad \forall v \in \mathcal{P}_{k+1}(K), &(2.51\text{b})\\ (u_h^\star, 1)_K &= (u_h, 1)_K. &(2.51\text{c})\end{aligned}$$

This postprocessing method was first proposed by Rolf Stenberg in [118]. A simple computation shows that for all $v \in \mathcal{P}_{k+1}(K)$

$$\begin{aligned}(\kappa\nabla u_h^\star, \nabla v)_K &= (f, v)_K - \langle \mathbf{q}_h\cdot\mathbf{n}, v\rangle_{\partial K}\\ &= (\mathrm{div}\,\mathbf{q}, v)_K - \langle \mathbf{q}_h\cdot\mathbf{n}, v\rangle_{\partial K}\\ &= -(\mathbf{q}, \nabla v)_K + \langle \mathbf{q}\cdot\mathbf{n} - \mathbf{q}_h\cdot\mathbf{n}, v\rangle_{\partial K}\\ &= (\kappa\nabla u, \nabla v)_K + \langle \mathbf{q}\cdot\mathbf{n} - \mathbf{q}_h\cdot\mathbf{n}, v\rangle_{\partial K}. \end{aligned}\tag{2.52}$$

Some preliminary comments. Before we start analyzing this, let us introduce the space

$$\mathcal{P}_{k+1}^0(K) := \{v \in \mathcal{P}_{k+1}(K) \,:\, (v, 1)_K = 0\},$$

and note that

$$(f, 1)_K = (\mathrm{div}\,\mathbf{q}_h, 1)_K = \langle \mathbf{q}_h\cdot\mathbf{n}, 1\rangle_{\partial K}, \tag{2.53}$$

which means that we can decompose in an orthogonal sum

$$u_h^\star = c_h + \omega_h, \qquad c_h \in \prod_{K\in\mathcal{T}_h} \mathcal{P}_0(K), \qquad \omega_h \in \prod_{K\in\mathcal{T}_h} \mathcal{P}_{k+1}^0(K), \tag{2.54a}$$

and compute separately for all $K \in \mathcal{T}_h$

$$\begin{aligned}(\kappa\nabla w_h, \nabla v)_K &= (f, v)_K - \langle \mathbf{q}_h\cdot\mathbf{n}, v\rangle_{\partial K} \qquad \forall v \in \mathcal{P}_{k+1}^0(K), &(2.54\text{b})\\ (c_h, 1)_K &= (u_h, 1)_K. &(2.54\text{c})\end{aligned}$$

It is clear that due to (2.53), problems (2.54) and (2.51) are equivalent, while it is quite obvious that problem (2.54) has a unique solution.

Lemma 2.3 *The following inequalities hold:*

$$\|v\|_{\partial K} \lesssim h_K^{1/2}|v|_{1,K}, \qquad \|v\|_K \approx h_K|v|_{1,K}, \qquad \forall v \in \mathcal{P}_{k+1}^0(K).$$

Proof Both inequalities follow from scaling arguments, and the following facts:

$$v \in \mathscr{P}^0_{k+1}(K) \quad \Longleftrightarrow \quad \widehat{v} \in \mathscr{P}^0_{k+1}(\widehat{K}), \tag{2.55}$$

$$\|v\|_{\partial\widehat{K}} \lesssim \|v\|_{\widehat{K}} \approx |v|_{1,\widehat{K}} \qquad \forall v \in \mathscr{P}^0_{k+1}(\widehat{K}). \tag{2.56}$$

It is also important to keep in mind (1.2), which says that the hat symbol is not ambiguous when applied in the interior domain or on the boundary. Then, the scaling argument is reduced to noticing that for all $v \in \mathscr{P}^0_{k+1}(K)$,

$$\begin{aligned}
\|v\|_{\partial K} &\approx h_K^{\frac{d-1}{2}} \|\widehat{v}\|_{\partial\widehat{K}} && \text{(scaling (1.7) and meaning of } \widehat{v})\\
&\lesssim h_K^{\frac{d-1}{2}} |\widehat{v}|_{1,\widehat{K}} && \text{(finite-dimensional bound (2.56))}\\
&\approx h_K^{\frac{1}{2}} |v|_{1,K}, && \text{(scaling (1.9))}
\end{aligned}$$

and

$$\begin{aligned}
\|v\|_K &\approx h_K^{\frac{d}{2}} \|\widehat{v}\|_{\widehat{K}} && \text{(scaling (1.7))}\\
&\approx h_K^{\frac{d}{2}} |\widehat{v}|_{1,\widehat{K}} && \text{(finite-dimensional bound (2.56))}\\
&\approx h_K |v|_{1,K}. && \text{(scaling (1.9))}
\end{aligned}$$

This completes the proof.

Proposition 2.6 (Postprocessing) *Let $(u_h, \mathbf{q}_h)$ be any approximation of the solution of* (2.44) *satisfying $(f, 1)_K = \langle \mathbf{q}_h \cdot \mathbf{n}, 1\rangle_{\partial K}$. Then the Stenberg postprocessing* (2.51) *satisfies*

$$\begin{aligned}
\|u - u_h^\star\|_\Omega \lesssim{}& \|u - \Pi_{k+1}u\|_\Omega + \Big(\sum_{K\in\mathscr{T}_h} h_K^2 |u - \Pi_{k+1}u|^2_{1,K}\Big)^{1/2}\\
&+ \|u_h - \Pi_k u\|_\Omega + h\|\mathbf{q}\cdot\mathbf{n} - \mathbf{q}_h\cdot\mathbf{n}\|_h.
\end{aligned}$$

If the discrete conservation property $(f, 1)_K = \langle \mathbf{q}_h \cdot \mathbf{n}, 1\rangle_{\partial K}$ does not hold, then the same bound is satisfied by the solution of (2.54).

Proof Let

$$\begin{aligned}
v &:= u_h^\star - \Pi_{k+1}u - \Pi_0(u_h^\star - \Pi_{k+1}u)\\
&= u_h^\star - \Pi_{k+1}u - \Pi_0(u_h - \Pi_k u), \quad \text{(by (2.51) and } \Pi_0\Pi_k = \Pi_0 = \Pi_0\Pi_{k+1})
\end{aligned}$$

and note that $v|_K \in \mathscr{P}^0_{k+1}(K)$. We then have

$$
\begin{aligned}
\|\kappa^{1/2}\nabla v\|_K^2 &= (\kappa\nabla(u_h^\star - \Pi_{k+1}u), \nabla v)_K && (\nabla \Pi_0 = 0)\\
&= (\kappa\nabla(u - \Pi_{k+1}u), \nabla v)_K + \langle \mathbf{q}\cdot\mathbf{n} - \mathbf{q}_h\cdot\mathbf{n}, v\rangle_{\partial K} && (\text{by (2.52)})\\
&\leqslant |u - \Pi_{k+1}u|_{1,K}\|\kappa^{1/2}\nabla v\|_K \|\kappa^{1/2}\|_{L^\infty}\\
&\quad + h_K^{1/2}\|\mathbf{q}\cdot\mathbf{n} - \mathbf{q}_h\cdot\mathbf{n}\|_{\partial K} h_K^{-1/2}\|v\|_{\partial K}\\
&\lesssim |u - \Pi_{k+1}u|_{1,K}\|\kappa^{1/2}\nabla v\|_K\\
&\quad + h_K^{1/2}\|\mathbf{q}\cdot\mathbf{n} - \mathbf{q}_h\cdot\mathbf{n}\|_{\partial K}\|\kappa^{1/2}\nabla v\|_K, && (\text{by Lemma 2.3})
\end{aligned}
$$

or, in other words,

$$
|v|_{1,K}^2 \lesssim |u - \Pi_{k+1}u|_{1,K}^2 + h_K\|\mathbf{q}\cdot\mathbf{n} - \mathbf{q}_h\cdot\mathbf{n}\|_{\partial K}^2. \tag{2.57}
$$

Therefore

$$
\begin{aligned}
&\|u_h^\star - \Pi_{k+1}u\|_K^2\\
&\quad = \|\Pi_0(u_h^\star - \Pi_{k+1}u)\|_K^2 + \|v\|_K^2 && (\text{orthogonal decomp})\\
&\quad = \|\Pi_0(u_h - \Pi_k u)\|_K^2 + \|v\|_K^2 && (\text{see definition of } v)\\
&\quad \lesssim \|u_h - \Pi_k u\|_K^2 + h_K^2|v|_{1,K}^2 && (\text{by Lemma 2.3})\\
&\quad \lesssim \|u_h - \Pi_k u\|_K^2\\
&\qquad + h_K^2|u - \Pi_{k+1}u|_{1,K}^2 + h_K^3\|\mathbf{q}\cdot\mathbf{n} - \mathbf{q}_h\cdot\mathbf{n}\|_{\partial K}^2, && (\text{by (2.57)})
\end{aligned}
$$

and to prove the result we only need to collect the contributions of all the elements.

For the RT discretization, assuming superconvergence, the Stenberg postprocessing (2.51) satisfies

$$
\|u - u_h^\star\|_\Omega \lesssim h^{k+2}.
$$

2.3.4 A Second Postprocessing Scheme

Another way of getting a good gradient. Since $\nabla u = -\kappa^{-1}\mathbf{q}$, we can use the approximation $\mathbf{q}_h$ as a way of getting an improved gradient, using u_h to determine the average of the postprocessed solution on each element. We then look for

$$
u_h^\star \in \prod_{K\in\mathcal{T}_h} \mathcal{P}_{k+1}(K), \tag{2.58a}
$$

satisfying for all $K \in \mathcal{T}_h$

$$
\begin{aligned}
(\nabla u_h^\star, \nabla v)_K &= -(\kappa^{-1}\mathbf{q}_h, \nabla v)_K \qquad \forall v \in \mathcal{P}_{k+1}^0(K), && (2.58b)\\
(u_h^\star, 1)_K &= (u_h, 1)_K. && (2.58c)
\end{aligned}
$$

Its analysis. Note that

$$(\nabla u_h^\star, \nabla v)_K = (\nabla u, \nabla v)_K + (\kappa^{-1}(\mathbf{q} - \mathbf{q}_h), \nabla v)_K \qquad \forall v \in \mathscr{P}_{k+1}^0(K).$$

Like in the proof of Proposition 2.6, by (2.58c) and $\Pi_0(\Pi_k - \Pi_{k+1}) = 0$, we have

$$\begin{aligned}\prod_{K\in\mathscr{T}_h} \mathscr{P}_{k+1}^0(K) \ni v &:= u_h^\star - \Pi_{k+1}u - \Pi_0(u_h^\star - \Pi_{k+1}u)\\ &= u_h^\star - \Pi_{k+1}u - \Pi_0(u_h - \Pi_k u).\end{aligned}$$

We then write

$$\begin{aligned}|v|_{1,K}^2 &= (\nabla(u_h^\star - \Pi_{k+1}u), \nabla v)_K\\ &= (\nabla(u - \Pi_{k+1}u), \nabla v)_K + (\kappa^{-1}(\mathbf{q} - \mathbf{q}_h), \nabla v)_K,\end{aligned}$$

so that, using Lemma 2.3, we have bounded

$$h_K^{-1}\|v\|_K \lesssim |v|_{1,K} \leqslant |u - \Pi_{k+1}u|_{1,K} + \|\kappa^{-1}(\mathbf{q} - \mathbf{q}_h)\|_K.$$

What is left follows the final steps of the arguments in Proposition 2.6, leading to

$$\begin{aligned}\|u - u_h^\star\|_\Omega \lesssim\ & \|u - \Pi_{k+1}u\|_\Omega + \|u_h - \Pi_k u\|_\Omega\\ &+ \left(\sum_{K\in\mathscr{T}_h} h_K^2 |u - \Pi_{k+1}u|_{1,K}^2\right)^{\frac{1}{2}} + h\|\mathbf{q} - \mathbf{q}_h\|_\Omega,\end{aligned}$$

and therefore to superconvergence. Once again, note that nothing particular about how $(\mathbf{q}_h, u_h)$ has been produced is used in this argument. However, to reach superconvergence, we obviously need that $\|u_h - \Pi_k u\|_\Omega$, superconverges, as is the case with the RT method.

2.3.5 The Influence of Reaction Terms

Reaction–diffusion problems. In this section, we will have a look at how the analysis of RT discretization is adapted for the following simple modification of our equations:

$$\kappa^{-1}\mathbf{q} + \nabla u = \mathbf{0} \quad \text{in } \Omega, \tag{2.59a}$$

$$\operatorname{div}\mathbf{q} + c\,u = f \quad \text{in } \Omega, \tag{2.59b}$$

$$u = g \quad \text{on } \Gamma, \tag{2.59c}$$

where $c : \Omega \to \mathbb{R}$ is a nonnegative function. The seminorm

$$|u|_c := (c\,u, u)_\Omega^{1/2} = \|c^{1/2} u\|_\Omega$$

will play an important role in the energy analysis of this problem.

Discretization and error equations. The RT equations for problem (2.59) are

$$\begin{aligned}
(\kappa^{-1}\mathbf{q}_h, \mathbf{r})_{\mathcal{T}_h} - (u_h, \operatorname{div}\mathbf{r})_{\mathcal{T}_h} + \langle \widehat{u}_h, \mathbf{r}\cdot\mathbf{n}\rangle_{\partial\mathcal{T}_h} &= 0 && \forall \mathbf{r} \in \mathbf{V}_h, && (2.60a)\\
(\operatorname{div}\mathbf{q}_h, w)_{\mathcal{T}_h} + (c\,u_h, w)_{\mathcal{T}_h} &= (f, w)_{\mathcal{T}_h} && \forall w \in W_h, && (2.60b)\\
\langle \mathbf{q}_h\cdot\mathbf{n}, \mu\rangle_{\partial\mathcal{T}_h\setminus\Gamma} &= 0 && \forall \mu \in M_h^\circ, && (2.60c)\\
\langle \widehat{u}_h, \mu\rangle_\Gamma &= \langle g, \mu\rangle_\Gamma && \forall \mu \in M_h^\Gamma, && (2.60d)
\end{aligned}$$

while projections satisfy the following discrete equations:

$$\begin{aligned}
(\kappa^{-1}\boldsymbol{\Pi}\mathbf{q}, \mathbf{r})_{\mathcal{T}_h} - (\Pi u, \operatorname{div}\mathbf{r})_{\mathcal{T}_h} + \langle \mathrm{P}u, \mathbf{r}\cdot\mathbf{n}\rangle_{\partial\mathcal{T}_h} &= (\kappa^{-1}\boldsymbol{\Pi}\mathbf{q} - \mathbf{q}, \mathbf{r})_{\mathcal{T}_h}, && (2.61a)\\
(\operatorname{div}\boldsymbol{\Pi}\mathbf{q}, w)_{\mathcal{T}_h} + (c\,\Pi u, w)_{\mathcal{T}_h} &= (f, w)_{\mathcal{T}_h}\\
&\quad + (c\,(\Pi u - u), w)_{\mathcal{T}_h}, && (2.61b)\\
\langle \boldsymbol{\Pi}\mathbf{q}\cdot\mathbf{n}, \mu_1\rangle_{\partial\mathcal{T}_h\setminus\Gamma} &= 0, && (2.61c)\\
\langle \mathrm{P}u, \mu_2\rangle_\Gamma &= \langle g, \mu_2\rangle_\Gamma, && (2.61d)
\end{aligned}$$

for all $(\mathbf{r}, w, \mu_1, \mu_2) \in \mathbf{V}_h \times W_h \times M_h^\circ \times M_h^\Gamma$. Subtracting the discrete Eqs. (2.60) from (2.61), we get the error equations

$$\begin{aligned}
(\kappa^{-1}\boldsymbol{\varepsilon}_h^q, \mathbf{r})_{\mathcal{T}_h} - (\varepsilon_h^u, \operatorname{div}\mathbf{r})_{\mathcal{T}_h} + \langle \widehat{\varepsilon}_h^u, \mathbf{r}\cdot\mathbf{n}\rangle_{\partial\mathcal{T}_h} &= (\kappa^{-1}(\boldsymbol{\Pi}\mathbf{q} - \mathbf{q}), \mathbf{r})_{\mathcal{T}_h}, && (2.62a)\\
(\operatorname{div}\boldsymbol{\varepsilon}_h^q, w)_{\mathcal{T}_h} + (c\,\varepsilon_h^u, w)_{\mathcal{T}_h} &= (c\,(\Pi u - u), w)_{\mathcal{T}_h}, && (2.62b)\\
\langle \boldsymbol{\varepsilon}_h^q\cdot\mathbf{n}, \mu_1\rangle_{\partial\mathcal{T}_h\setminus\Gamma} &= 0, && (2.62c)\\
\langle \widehat{\varepsilon}_h^u, \mu_2\rangle_\Gamma &= 0. && (2.62d)
\end{aligned}$$

Testing the error Eqs. (2.62) with $(\boldsymbol{\varepsilon}_h^q, \varepsilon_h^u, -\widehat{\varepsilon}_h^u, -\boldsymbol{\varepsilon}_h^q\cdot\mathbf{n})$ and adding the result, we reach the new energy identity

$$(\kappa^{-1}\boldsymbol{\varepsilon}_h^q, \boldsymbol{\varepsilon}_h^q)_{\mathcal{T}_h} + (c\,\varepsilon_h^u, \varepsilon_h^u)_{\mathcal{T}_h} = (\kappa^{-1}(\boldsymbol{\Pi}\mathbf{q} - \mathbf{q}), \boldsymbol{\varepsilon}_h^q)_{\mathcal{T}_h} + (c\,(\Pi u - u), \varepsilon_h^u)_{\mathcal{T}_h},$$

thus proving the estimate

$$\|\boldsymbol{\varepsilon}_h^q\|_{\kappa^{-1}}^2 + |\varepsilon_h^u|_c^2 \leqslant \|\boldsymbol{\Pi}\mathbf{q} - \mathbf{q}\|_{\kappa^{-1}}^2 + |\Pi u - u|_c^2. \tag{2.63}$$

As can be seen from (2.63), this couples back the estimates for the variable $\mathbf{q}$ with the approximation properties of W_h. The estimate (see (2.34))

$$\|\boldsymbol{\varepsilon}_h^q\cdot\mathbf{n}\|_h \lesssim \|\boldsymbol{\varepsilon}_h^q\|_\Omega \tag{2.64}$$

is a purely finite-dimensional one, independent of the equations satisfied by the discrete quantities. In a similar spirit, we can prove (2.40) again, i.e., we obtain

$$\|\mathrm{P}u - \widehat{u}_h\|_h = \|\widehat{\varepsilon}_h^u\|_h \lesssim \|\varepsilon_h^u\|_\Omega + h\|\mathbf{q} - \mathbf{q}_h\|_{\kappa^{-1}}. \tag{2.65}$$

This happens because this estimate depends only on the first error equation (the discretization of the equation $\kappa^{-1}\mathbf{q} + \nabla u = \mathbf{0}$) which does not depend on the particular equilibrium equation. The argument to prove that

$$\|\Pi u - u_h\|_\Omega = \|\varepsilon_h^u\|_\Omega \lesssim \|\mathbf{q} - \mathbf{q}_h\|_{\kappa^{-1}} \tag{2.66}$$

was based on the commutativity property for the projection and on the first error equation, so nothing has to be changed.

The duality estimate. The duality argument becomes more complicated as we deal with more complex model problems. Instead of adapting the proof of the superconvergence estimate of the diffusion problem, we are going to show a more systematic way of proving estimates, a methodology that will be extremely useful in HDG analysis. We start with the dual problem

$$\kappa^{-1}\boldsymbol{\xi} - \nabla\theta = \mathbf{0} \qquad \text{in } \Omega, \tag{2.67a}$$
$$-\mathrm{div}\,\boldsymbol{\xi} + c\,\theta = \varepsilon_h^u \qquad \text{in } \Omega, \tag{2.67b}$$
$$\theta = 0 \qquad \text{on } \Gamma. \tag{2.67c}$$

Note that this time we have changed signs in both first-order operators. We assume a regularity hypothesis

$$\|\boldsymbol{\xi}\|_{1,\Omega} + \|\theta\|_{1,\Omega} + \|c\,\theta\|_{1,\Omega} \leqslant C_{\mathrm{reg}}\|\varepsilon_h^u\|_\Omega. \tag{2.68}$$

We start by writing down the discrete equations satisfied by the projections $(\boldsymbol{\Pi}\boldsymbol{\xi}, \Pi\theta, \mathrm{P}\theta)$:

$$(\kappa^{-1}\boldsymbol{\Pi}\boldsymbol{\xi}, \mathbf{r})_{\mathcal{T}_h} + (\Pi\theta, \mathrm{div}\,\mathbf{r})_{\mathcal{T}_h} - \langle \mathrm{P}\theta, \mathbf{r}\cdot\mathbf{n}\rangle_{\partial\mathcal{T}_h} = (\kappa^{-1}\boldsymbol{\Pi}\boldsymbol{\xi} - \boldsymbol{\xi}, \mathbf{r})_{\mathcal{T}_h}, \tag{2.69a}$$
$$-(\mathrm{div}\,\boldsymbol{\Pi}\boldsymbol{\xi}, w)_{\mathcal{T}_h} + (c\,\Pi\theta, w)_{\mathcal{T}_h} = (\varepsilon_h^u, w)_{\mathcal{T}_h} + (c\,(\Pi\theta - \theta), w)_{\mathcal{T}_h}, \tag{2.69b}$$
$$\langle \boldsymbol{\Pi}\boldsymbol{\xi}\cdot\mathbf{n}, \mu_1\rangle_{\partial\mathcal{T}_h\setminus\Gamma} = 0, \tag{2.69c}$$
$$\langle \mathrm{P}\theta, \mu_2\rangle_\Gamma = \langle u_0, \mu_2\rangle_\Gamma, \tag{2.69d}$$

for all $(\mathbf{r}, w, \mu_1, \mu_2) \in \mathbf{V}_h \times W_h \times M_h^\circ \times M_h^\Gamma$. We now test the first three equations with $(\boldsymbol{\varepsilon}_h^q, \varepsilon_h^u, \widehat{\varepsilon}_h^u)$ and align terms carefully:

$$(\boldsymbol{\Pi\xi}, \kappa^{-1}\boldsymbol{\varepsilon}_h^q)_{\mathcal{T}_h} + (\Pi\theta, \operatorname{div}\boldsymbol{\varepsilon}_h^q)_{\mathcal{T}_h} - \langle \mathrm{P}\theta, \widehat{\varepsilon}_h^u\rangle_{\partial\mathcal{T}_h} = (\boldsymbol{\Pi\xi} - \boldsymbol{\xi}, \kappa^{-1}\varepsilon_h^q)_{\mathcal{T}_h}, \tag{2.70a}$$

$$-(\operatorname{div}\boldsymbol{\Pi\xi}, \varepsilon_h^u)_{\mathcal{T}_h} + (\Pi\theta, c\,\varepsilon_h^u)_{\mathcal{T}_h} = \|\varepsilon_h^u\|_\Omega^2 + (\Pi\theta - \theta, c\,\varepsilon_h^u)_{\mathcal{T}_h}, \tag{2.70b}$$

$$\langle \boldsymbol{\Pi\xi}\cdot\mathbf{n}, \widehat{\varepsilon}_h^u\rangle_{\partial\mathcal{T}_h} = 0. \tag{2.70c}$$

Note that we have used twice that $\widehat{\varepsilon}_h^u = 0$ on Γ (this is the fourth of the error Eqs. (2.62)). The next course of action is the addition of Eqs. (2.70). Close inspection of the columns of the tabulated system (2.70) shows the error Eqs. (2.62) tested with $(\boldsymbol{\Pi\xi}, \Pi\theta, \mathrm{P}\theta)$. Therefore

$$\begin{aligned}&(\kappa^{-1}(\boldsymbol{\Pi}\mathbf{q} - \mathbf{q}), \boldsymbol{\Pi\xi})_{\mathcal{T}_h} + (c(\Pi u - u), \Pi\theta)_{\mathcal{T}_h}\\ &\quad = \|\varepsilon_h^u\|_\Omega^2 + (c\,(\Pi u - u_h), \Pi\theta - \theta)_{\mathcal{T}_h} + (\kappa^{-1}(\boldsymbol{\Pi}\mathbf{q} - \mathbf{q}_h), \boldsymbol{\Pi\xi} - \boldsymbol{\xi})_{\mathcal{T}_h}.\end{aligned}$$

We just reorganize this equality to get

$$\begin{aligned}\|\varepsilon_h^u\|_\Omega^2 &= (\kappa^{-1}(\boldsymbol{\Pi}\mathbf{q} - \mathbf{q}), \boldsymbol{\Pi\xi})_{\mathcal{T}_h} - (\kappa^{-1}(\boldsymbol{\Pi}\mathbf{q} - \mathbf{q}_h), \boldsymbol{\Pi\xi} - \boldsymbol{\xi})_{\mathcal{T}_h}\\ &\quad + (c(\Pi u - u), \Pi\theta)_{\mathcal{T}_h} - (c\,(\Pi u - u_h), \Pi\theta - \theta)_{\mathcal{T}_h}\\ &= (\kappa^{-1}(\mathbf{q}_h - \mathbf{q}), \boldsymbol{\Pi\xi} - \boldsymbol{\xi})_{\mathcal{T}_h} + (\kappa^{-1}(\boldsymbol{\Pi}\mathbf{q} - \mathbf{q}), \boldsymbol{\xi})_{\mathcal{T}_h} && (\pm\boldsymbol{\xi})\\ &\quad + (c\,(u_h - u), \Pi\theta - \theta)_{\mathcal{T}_h} + (c\,(\Pi u - u), \theta)_{\mathcal{T}_h} && (\pm\theta)\\ &= (\kappa^{-1}(\mathbf{q}_h - \mathbf{q}), \boldsymbol{\Pi\xi} - \boldsymbol{\xi})_{\mathcal{T}_h} + (c\,(u_h - u), \Pi\theta - \theta)_{\mathcal{T}_h}\\ &\quad + (\boldsymbol{\Pi}\mathbf{q} - \mathbf{q}, \nabla\theta)_{\mathcal{T}_h} + (\Pi u - u, c\,\theta)_{\mathcal{T}_h}. && (\kappa^{-1}\boldsymbol{\xi} = \nabla\theta)\end{aligned}$$

The last two terms need some additional work. The second one (of the last two terms) is easy:

$$(\Pi u - u, c\,\theta)_{\mathcal{T}_h} = (\Pi u - u, c\theta - \Pi(c\,\theta))_{\mathcal{T}_h}.$$

In the first one, we start with integration by parts

$$\begin{aligned}&(\boldsymbol{\Pi}\mathbf{q} - \mathbf{q}, \nabla\theta)_\Omega\\ &\quad = -(\operatorname{div}(\boldsymbol{\Pi}\mathbf{q} - \mathbf{q}), \theta)_\Omega + \langle(\boldsymbol{\Pi}\mathbf{q} - \mathbf{q})\cdot\mathbf{n}, \theta\rangle_\Gamma && (\boldsymbol{\Pi}\mathbf{q} - \mathbf{q} \in \mathbf{H}(\mathrm{div}, \Omega))\\ &\quad = -(\operatorname{div}(\boldsymbol{\Pi}\mathbf{q} - \mathbf{q}), \theta)_\Omega && (\text{BC for dual problem})\\ &\quad = -(\Pi(\operatorname{div}\mathbf{q}) - \operatorname{div}\mathbf{q}, \theta)_\Omega && (\text{commutativity prop.})\\ &\quad = (\Pi(\operatorname{div}\mathbf{q}) - \operatorname{div}\mathbf{q}, \Pi\theta - \theta)_\Omega.\end{aligned}$$

Collecting these equalities and applying bounds on low-order estimates for the projections, we get

$$\begin{aligned}\|\varepsilon_h^u\|_\Omega^2 &\lesssim h\|\mathbf{q} - \mathbf{q}_h\|_\Omega\,|\boldsymbol{\xi}|_{1,\Omega} + h|u - u_h|_c|\theta|_{1,\Omega}\\ &\quad + h\|\Pi(\operatorname{div}\mathbf{q}) - \operatorname{div}\mathbf{q}\|_\Omega|\theta|_{1,\Omega} + h\|\Pi u - u\|_\Omega|c\,\theta|_{1,\Omega},\end{aligned}$$

which together with the regularity assumption (2.68) and the energy estimate (2.63) proves superconvergence:

$$\|\varepsilon_h^u\|_\Omega \lesssim h(\|\boldsymbol{\Pi}\mathbf{q} - \mathbf{q}\|_\Omega + \|\Pi u - u\|_\Omega + \|\Pi(\operatorname{div}\mathbf{q}) - \operatorname{div}\mathbf{q}\|_\Omega). \tag{2.71}$$

Some notes. As can be seen from these arguments, duality estimates are not a smooth ride, but they follow quite predictable patterns. The reader can wonder how it was the case that the duality estimate when $c = 0$ seemed so much simpler. There is a simple reason: when $c = 0$, then $\operatorname{div}\boldsymbol{\varepsilon}_h^q = 0$, and it is simple to show (from the first error equation) that

$$(\kappa^{-1}\boldsymbol{\xi}, \boldsymbol{\varepsilon}_h^q)_{\mathscr{T}_h} = 0,$$

which takes us back to some of the simpler estimates of Sect. 2.2.4. Note also that when $k \geqslant 1$, we can write

$$(\boldsymbol{\Pi}\mathbf{q} - \mathbf{q}, \nabla\theta)_\Omega = (\boldsymbol{\Pi}\mathbf{q} - \mathbf{q}, \nabla\theta - \boldsymbol{\Pi}_0(\nabla\theta))_\Omega \lesssim h\|\boldsymbol{\Pi}\mathbf{q} - \mathbf{q}\|_\Omega |\theta|_{2,\Omega},$$

which leads to a slightly different regularity assumption and does not require integration by parts to make the additional power of h. This argument does not hold in the lower order case $k = 0$, because the projection does not include any internal degrees of freedom.

2.4 Introducing BDM

In this section, we go over all the needed changes to modify the projection-based analysis of RT elements to a similar analysis of a loosely called Brezzi–Douglas–Marini BDM element (we'll deal with names later on). For purposes of comparison, we will stick to the following table, lining up the boundary d.o.f. and not the space that we used for the variable u_h. (The definition of the Nédélec space $\mathscr{N}_{k-2}$ is given in Sect. 2.4.1.)

degree	$\mathbf{q}_h$	u_h	bd. d.o.f.	int. d.o.f.
$k \geqslant 0$	$\mathscr{RT}_k(K) = \boldsymbol{\mathscr{P}}_k(K) + \mathbf{m}\widetilde{\mathscr{P}}_k(K)$	$\mathscr{P}_k(K)$	$\mathscr{R}_k(\partial K)$	$\boldsymbol{\mathscr{P}}_{k-1}(K)$
$k \geqslant 1$	$\boldsymbol{\mathscr{P}}_k(K)$	$\mathscr{P}_{k-1}(K)$	$\mathscr{R}_k(\partial K)$	$\mathscr{N}_{k-2}(K)$

2.4.1 The Nédélec Space

Consider the spaces

$$\mathscr{N}_k(K) := \boldsymbol{\mathscr{P}}_k(K) \oplus \{\mathbf{q} \in \widetilde{\boldsymbol{\mathscr{P}}}_{k+1}(K) : \mathbf{q} \cdot \mathbf{m} = 0\},$$

which obviously satisfy

$$\boldsymbol{\mathscr{P}}_k(K) \subset \mathscr{N}_k(K) \subset \boldsymbol{\mathscr{P}}_{k+1}(K).$$

Proposition 2.7 *The following properties hold:*

(a) $\dim \mathscr{N}_k(K) = d \dim \mathscr{P}_{k+1}(K) - \dim \widetilde{\mathscr{P}}_{k+2}(K)$.
(b) $\dim \mathscr{N}_{k-1}(K) + \dim \mathscr{R}_{k+1}(\partial K) = \dim \boldsymbol{\mathscr{P}}_{k+1}(K)$.
(c) $\mathscr{N}_k(K) \oplus \nabla \widetilde{\mathscr{P}}_{k+2}(K) = \boldsymbol{\mathscr{P}}_{k+1}(K)$.
(d) $\mathbf{q} \in \mathscr{N}_k(K) \iff \check{\mathbf{q}} \in \mathscr{N}_k(\widehat{K})$.

Proof It is easy to see that the linear operator

$$\widetilde{\boldsymbol{\mathscr{P}}}_{k+1}(K) \ni \mathbf{p} \longmapsto T\mathbf{p} := \mathbf{p} \cdot \mathbf{m} \in \widetilde{\mathscr{P}}_{k+2}(K)$$

is onto. Hence,

$$\begin{aligned}
&\dim \mathscr{N}_k(K) \\
&\quad = \dim \boldsymbol{\mathscr{P}}_k(K) + \dim \ker T && (\mathscr{N}_k = \boldsymbol{\mathscr{P}}_k \oplus \ker T) \\
&\quad = d \dim \mathscr{P}_k(K) + \dim \widetilde{\boldsymbol{\mathscr{P}}}_{k+1}(K) - \dim \widetilde{\mathscr{P}}_{k+2}(K) && (T \text{ is onto}) \\
&\quad = d(\dim \mathscr{P}_k(K) + \dim \widetilde{\mathscr{P}}_{k+1}(K)) - \dim \widetilde{\mathscr{P}}_{k+2}(K),
\end{aligned}$$

which proves (a). To prove (b), note that by (a)

$$\begin{aligned}
\dim \mathscr{N}_{k-1}(K) + \dim \mathscr{R}_{k+1}(K) &= d \dim \mathscr{P}_k(K) - \dim \widetilde{\mathscr{P}}_{k+1}(K) \\
&\quad + (d+1)\dim \mathscr{P}_{k+1}(e) \\
&= d \dim \mathscr{P}_k(K) + d \dim \widetilde{\mathscr{P}}_{k+1}(K),
\end{aligned}$$

where e denotes any of the faces of K. To prove (c), we set

$$\mathscr{S}_{k+1} = \{\mathbf{q} \in \widetilde{\boldsymbol{\mathscr{P}}}_{k+1}(K) : \mathbf{q} \cdot \mathbf{m} = 0\}.$$

On the one hand $\mathscr{S}_{k+1} + \nabla \widetilde{\mathscr{P}}_{k+2} \subseteq \widetilde{\boldsymbol{\mathscr{P}}}_{k+1}$ and this sum is direct, since if $p \in \widetilde{\mathscr{P}}_{k+2}$, then $\nabla p \cdot \mathbf{m} = (k+2)p$ by the Euler homogeneous function theorem. Finally,

$$\begin{aligned}
&\dim (\mathscr{S}_{k+1} \oplus \nabla \widetilde{\mathscr{P}}_{k+2}) \\
&\quad = \dim \mathscr{S}_{k+1} + \dim \widetilde{\mathscr{P}}_{k+2} && (\nabla \text{ is 1-1}) \\
&\quad = d \dim \widetilde{\mathscr{P}}_{k+1} - \dim \widetilde{\mathscr{P}}_{k+2} + \dim \widetilde{\mathscr{P}}_{k+2} && (\text{computation to prove (a)}) \\
&\quad = \dim \widetilde{\boldsymbol{\mathscr{P}}}_{k+1},
\end{aligned}$$

and therefore $\mathscr{S}_{k+1} \oplus \nabla \widetilde{\mathscr{P}}_{k+2} = \widetilde{\boldsymbol{\mathscr{P}}}_{k+1}$ and (c) holds.

To prove (d), we just need to show that if $\mathbf{q} \in \mathscr{S}_{k+1}$, then $\check{\mathbf{q}} \in \mathscr{N}_k(\widehat{K})$ (note that the transformation $\mathbf{q} \mapsto \check{\mathbf{q}}$ is a bijection). Take $\mathbf{q} \in \mathscr{S}_{k+1}$ and $\mathrm{F}_K(\widehat{\mathbf{x}}) = \mathrm{B}_K\widehat{\mathbf{x}} + \mathbf{b}_K$ to see

$$\begin{aligned}\check{\mathbf{q}}(\widehat{\mathbf{x}}) \cdot (\widehat{\mathbf{x}} + \mathrm{B}_K^{-1}\mathbf{b}_K) &= \mathrm{B}_K^\top \mathbf{q}(\mathrm{F}_K(\widehat{\mathbf{x}})) \cdot (\widehat{\mathbf{x}} + \mathrm{B}_K^{-1}\mathbf{b}_K) && \text{(definition of } \check{\mathbf{q}}) \\ &= \mathbf{q}(\mathrm{F}_K(\widehat{\mathbf{x}})) \cdot (\mathrm{B}_K\widehat{\mathbf{x}} + \mathbf{b}_K) \\ &= \mathbf{q}(\mathrm{F}_K(\widehat{\mathbf{x}})) \cdot \mathrm{F}_K(\widehat{\mathbf{x}}) = 0. && \text{(since } \mathbf{q} \in \mathscr{S}_{k+1})\end{aligned}$$

If we now decompose $\check{\mathbf{q}} \in \boldsymbol{\mathscr{P}}_{k+1}(\widehat{K}) = \widetilde{\mathbf{q}} + \mathbf{q}_k$, where $\widetilde{\mathbf{q}} \in \widetilde{\boldsymbol{\mathscr{P}}}_{k+1}(\widehat{K})$ and $\mathbf{q}_k \in \boldsymbol{\mathscr{P}}_k(\widehat{K})$, then we have (with $\mathbf{c} := \mathrm{B}_K^{-1}\mathbf{b}_K$)

$$0 = \check{\mathbf{q}}(\widehat{\mathbf{x}}) \cdot (\widehat{\mathbf{x}} + \mathbf{c}) = \underbrace{\widetilde{\mathbf{q}}(\widehat{\mathbf{x}}) \cdot \widehat{\mathbf{x}}}_{\in \widetilde{\mathscr{P}}_{k+2}(\widehat{K})} + \underbrace{\widetilde{\mathbf{q}}(\widehat{\mathbf{x}}) \cdot \mathbf{c} + \mathbf{q}_k(\widehat{\mathbf{x}}) \cdot (\widehat{\mathbf{x}} + \mathbf{c})}_{\in \mathscr{P}_{k+1}(\widehat{K})},$$

and therefore $\widetilde{\mathbf{q}} \cdot \mathbf{m} = 0$, which implies that $\widetilde{\mathbf{q}} \in \mathscr{S}_{k+1}$ and therefore $\check{\mathbf{q}} \in \mathscr{N}_k(\widehat{K})$.

The two-dimensional spaces. When $d = 2$, it is easy to see that

$$(q_1, q_2) \in \mathscr{R}\mathscr{T}_k(K) \quad \Longrightarrow \quad (-q_2, q_1) \in \mathscr{N}_k(K),$$

and $\dim \mathscr{R}\mathscr{T}_k(K) = \dim \mathscr{N}_k(K)$. Therefore

$$(q_1, q_2) \in \mathscr{R}\mathscr{T}_k(K) \quad \Longleftrightarrow \quad (-q_2, q_1) \in \mathscr{N}_k(K),$$

which means that the Nédélec space follows from a $\pi/2$-rotation of the Raviart–Thomas space in two space dimensions.

2.4.2 The BDM Projection

The BDM projection is the interpolation operator associated to a mixed finite element named after Brezzi, Douglas, and Marini. The BDM element was first introduced in the two-dimensional case [9], with slightly different internal degrees of freedom from those we are going to see here. The three-dimensional version that we will see here is due to Nédélec [93]. There is another variant of this three-dimensional BDM element due to Brezzi et al. [8].

The BDM projection. Let $\mathbf{q} : K \to \mathbb{R}^d$ be sufficiently smooth. For $k \geqslant 1$, the BDM projection is $\boldsymbol{\Pi}^{\mathrm{BDM}}\mathbf{q} \in \boldsymbol{\mathscr{P}}_k(K)$ characterized by the equations

$$(\boldsymbol{\Pi}^{\mathrm{BDM}}\mathbf{q}, \mathbf{r})_K = (\mathbf{q}, \mathbf{r})_K \qquad \forall \mathbf{r} \in \mathscr{N}_{k-2}(K), \tag{2.72a}$$

$$\langle \boldsymbol{\Pi}^{\mathrm{BDM}}\mathbf{q} \cdot \mathbf{n}, \mu \rangle_{\partial K} = \langle \mathbf{q} \cdot \mathbf{n}, \mu \rangle_{\partial K} \qquad \forall \mu \in \mathscr{R}_k(\partial K). \tag{2.72b}$$

The associated scalar projection is $\Pi_{k-1} u \in \mathscr{P}_{k-1}(K)$

$$(\Pi_{k-1} u, v)_K = (u, v)_K \qquad \forall v \in \mathscr{P}_{k-1}(K). \tag{2.72c}$$

In the case $k = 1$, Eqs. (2.72a) are void.

Proposition 2.8 (Definition of the BDM projection) *Equations* (2.72a) *and* (2.72b) *are uniquely solvable.*

Proof By Proposition 2.7(b), these equations make up a square system of linear equations, so we only need to prove uniqueness of solution of the homogeneous problem. Let thus $\mathbf{q} \in \boldsymbol{\mathscr{P}}_k(K)$ satisfy

$$(\mathbf{q}, \mathbf{r})_K = 0 \qquad \forall \mathbf{r} \in \mathscr{N}_{k-2}(K), \tag{2.73a}$$

$$\langle \mathbf{q} \cdot \mathbf{n}, \mu \rangle_{\partial K} = 0 \qquad \forall \mu \in \mathscr{R}_k(\partial K). \tag{2.73b}$$

Equation (2.73b) implies that $\mathbf{q} \cdot \mathbf{n} = 0$ on ∂K. Now take $u \in \widetilde{\mathscr{P}}_k(K)$ and note that

$$\begin{aligned}
(\mathbf{q}, \nabla u)_K &= -(\operatorname{div} \mathbf{q}, u)_K && \text{(integration by parts and } \mathbf{q} \cdot \mathbf{n} = 0) \\
&= -(\operatorname{div} \mathbf{q}, \Pi_{k-1} u)_K && (\operatorname{div} \mathbf{q} \in \mathscr{P}_{k-1}(K)) \\
&= (\mathbf{q}, \nabla \Pi_{k-1} u)_K && \text{(integration by parts and } \mathbf{q} \cdot \mathbf{n} = 0) \\
&= 0. && (\nabla \Pi_{k-1} u \in \boldsymbol{\mathscr{P}}_{k-2}(K) \subset \mathscr{N}_{k-2}(K) \text{ and (2.73a))}
\end{aligned}$$

Therefore $(\mathbf{q}, \mathbf{r})_K = 0$ for all $\mathbf{r} \in \mathscr{N}_{k-2}(K) + \nabla \widetilde{\mathscr{P}}_k(K) = \boldsymbol{\mathscr{P}}_{k-1}(K)$ (this was proved in Proposition 2.7(c)). This means that $\mathbf{q} \in \boldsymbol{\mathscr{P}}_k^{\perp}(K)$ and $\mathbf{q} \cdot \mathbf{n} = 0$ on ∂K, which implies (by Lemma 2.1(b)) that $\mathbf{q} = \mathbf{0}$.

The commutativity property. For all $\mathbf{q}$ and $u \in \mathscr{P}_{k-1}(K)$, we have

$$\begin{aligned}
&(\operatorname{div} \boldsymbol{\Pi}^{\mathrm{BDM}} \mathbf{q}, u)_K \\
&= \langle \boldsymbol{\Pi}^{\mathrm{BDM}} \mathbf{q} \cdot \mathbf{n}, u \rangle_{\partial K} - (\boldsymbol{\Pi}^{\mathrm{BDM}} \mathbf{q}, \nabla u)_K \\
&= \langle \mathbf{q} \cdot \mathbf{n}, u \rangle_{\partial K} - (\mathbf{q}, \nabla u)_K && \text{(by (2.72) and } \nabla u \in \mathscr{N}_{k-2}(K)) \\
&= (\operatorname{div} \mathbf{q}, u)_K,
\end{aligned}$$

and therefore

$$\operatorname{div} \boldsymbol{\Pi}^{\mathrm{BDM}} \mathbf{q} = \Pi_{k-1}(\operatorname{div} \mathbf{q}). \tag{2.74}$$

Invariance by Piola transforms. Let $\widehat{\boldsymbol{\Pi}}^{\mathrm{BDM}}$ be the BDM projection on the reference triangle. Using the formulas for change to the reference domain, we have

$$(\widehat{\boldsymbol{\Pi}^{\mathrm{BDM}} \mathbf{q}}, \check{\mathbf{r}})_{\widehat{K}} = (\boldsymbol{\Pi}^{\mathrm{BDM}} \mathbf{q}, \mathbf{r})_K = (\mathbf{q}, \mathbf{r})_K = (\widehat{\mathbf{q}}, \check{\mathbf{r}})_{\widehat{K}} \qquad \forall \mathbf{r} \in \mathscr{N}_{k-2}(K),$$

(see Proposition 2.7(d)), and

$$\begin{aligned}
\langle \widehat{\boldsymbol{\Pi}^{\mathrm{BDM}} \mathbf{q}} \cdot \widehat{\mathbf{n}}, \widehat{\mu} \rangle_{\partial \widehat{K}} &= \langle \boldsymbol{\Pi}^{\mathrm{BDM}} \mathbf{q} \cdot \mathbf{n}, \mu \rangle_{\partial K} \\
&= \langle \mathbf{q} \cdot \mathbf{n}, \mu \rangle_{\partial K} = \langle \widehat{\mathbf{q}} \cdot \widehat{\mathbf{n}}, \widehat{\mu} \rangle_{\partial \widehat{K}} \qquad \forall \mu \in \mathscr{R}_k(\partial K),
\end{aligned}$$

which proves that

$$\widehat{\boldsymbol{\Pi}^{\mathrm{BDM}}\mathbf{q}} = \widehat{\boldsymbol{\Pi}}^{\mathrm{BDM}}\widehat{\mathbf{q}}. \tag{2.75}$$

Proposition 2.9 (Estimates for the BDM projection) *On shape-regular triangulations and for sufficiently smooth* $\mathbf{q}$,

(a) $\|\boldsymbol{\Pi}^{\mathrm{BDM}}\mathbf{q}\|_K \lesssim \|\mathbf{q}\|_K + h_K|\mathbf{q}|_{1,K}$,
(b) $\|\mathbf{q} - \boldsymbol{\Pi}^{\mathrm{BDM}}\mathbf{q}\|_K \lesssim h_K^{k+1}|\mathbf{q}|_{k+1,K}$,
(c) $\|\mathrm{div}\,\mathbf{q} - \mathrm{div}\,\boldsymbol{\Pi}^{\mathrm{BDM}}\mathbf{q}\|_K \lesssim h_K^k|\mathrm{div}\,\mathbf{q}|_{k,K}$.

Proof The proof is almost identical to that of Proposition 2.3. We first need to show that

$$\|\widehat{\boldsymbol{\Pi}}^{\mathrm{BDM}}\widehat{\mathbf{q}}\|_{\widehat{K}} \lesssim \|\widehat{\mathbf{q}}\|_{\widehat{K}} + \|\widehat{\mathbf{q}}\cdot\widehat{\mathbf{n}}\|_{\partial\widehat{K}} \lesssim \|\widehat{\mathbf{q}}\|_{1,\widehat{K}} \qquad \forall \widehat{\mathbf{q}} \in \mathbf{H}^1(\widehat{K}),$$

and then use a scaling argument, taking advantage of (2.75), to move to the reference element, to prove (a). Since $\widehat{\Pi}^{\mathrm{BDM}}$ preserves the space $\boldsymbol{\mathcal{P}}_k$, we have

$$\|\widehat{\mathbf{q}} - \widehat{\boldsymbol{\Pi}}^{\mathrm{BDM}}\widehat{\mathbf{q}}\|_{\widehat{K}} \lesssim |\widehat{\mathbf{q}}|_{k+1,\widehat{K}} \qquad \forall \widehat{\mathbf{q}} \in \mathbf{H}^{k+1}(\widehat{K}).$$

Another scaling argument then proves (b). (Note that the details of these scaling arguments are the same as in Proposition 2.3.) Finally (c) follows from (2.74).

Proposition 2.10 (BDM local lifting of the normal trace) *For* $k \geqslant 1$, *there exists a linear operator* $\mathbf{L}^{\mathrm{BDM}} : \mathcal{R}_k(\partial K) \to \boldsymbol{\mathcal{P}}_k(K)$ *such that*

$$(\mathbf{L}^{\mathrm{BDM}}\mu)\cdot\mathbf{n} = \mu \quad \textit{and} \quad \|\mathbf{L}^{\mathrm{BDM}}\mu\|_K \lesssim h_K^{1/2}\|\mu\|_{\partial K} \qquad \forall \mu \in \mathcal{R}_k(\partial K).$$

Proof Let $\mathbf{q} = \mathbf{L}^{\mathrm{BDM}}\mu \in \boldsymbol{\mathcal{P}}_k(K)$ be defined as $\mathbf{q} = |J_K|^{-1}\mathrm{B}_K\widehat{\mathbf{q}} \circ \mathrm{G}_K$, where $\widehat{\mathbf{q}} \in \boldsymbol{\mathcal{P}}_k(\widehat{K})$ is the solution of the discrete equations in the reference domain:

$$\begin{aligned} (\widehat{\mathbf{q}}, \mathbf{r})_{\widehat{K}} &= 0 && \forall \mathbf{r} \in \mathcal{N}_{k-2}(\widehat{K}), \\ \langle \widehat{\mathbf{q}}\cdot\widehat{\mathbf{n}}, \xi\rangle_{\partial\widehat{K}} &= \langle \check{\mu}, \xi\rangle_{\partial\widehat{K}} && \forall \xi \in \mathcal{R}_k(\partial\widehat{K}). \end{aligned}$$

The remainder of the proof of Proposition 2.4 (essentially a scaling argument) can be used word by word to prove the result.

2.4.3 The BDM Method

Spaces and equations. We start by redefining the discrete spaces

$$\mathbf{V}_h := \prod_{K\in\mathcal{T}_h} \boldsymbol{\mathcal{P}}_k(K), \qquad W_h := \prod_{K\in\mathcal{T}_h} \mathcal{P}_{k-1}(K), \qquad M_h := \prod_{e\in\mathcal{E}_h} \mathcal{P}_k(e),$$

and similarly M_h° and M_h^Γ. We look for

$$(\mathbf{q}_h, u_h, \widehat{u}_h) \in \mathbf{V}_h \times W_h \times M_h, \tag{2.76a}$$

satisfying

$$\begin{aligned}
(\kappa^{-1}\mathbf{q}_h, \mathbf{r})_{\mathscr{T}_h} - (u_h, \operatorname{div}\mathbf{r})_{\mathscr{T}_h} + \langle \widehat{u}_h, \mathbf{r}\cdot\mathbf{n}\rangle_{\partial\mathscr{T}_h} &= 0 && \forall \mathbf{r} \in \mathbf{V}_h, && (2.76b)\\
(\operatorname{div}\mathbf{q}_h, w)_{\mathscr{T}_h} &= (f, w)_{\mathscr{T}_h} && \forall w \in W_h, && (2.76c)\\
\langle \mathbf{q}_h\cdot\mathbf{n}, \mu\rangle_{\partial\mathscr{T}_h\setminus\Gamma} &= 0 && \forall \mu \in M_h^\circ, && (2.76d)\\
\langle \widehat{u}_h, \mu\rangle_\Gamma &= \langle u_0, \mu\rangle_\Gamma && \forall \mu \in M_h^\Gamma. && (2.76e)
\end{aligned}$$

A reduced conforming formulation, involving $\mathbf{q}_h$ and u_h only, can be obtained using $\mathbf{V}_h^{\mathrm{div}} = \mathbf{V}_h \cap \mathbf{H}(\mathrm{div}, \Omega)$ as the test space in (2.76b) and noticing that (2.76d) is equivalent to $\mathbf{q}_h \in \mathbf{V}_h^{\mathrm{div}}$.

Proposition 2.11 *Equations* (2.76) *have a unique solution.*

Proof (This proof is a simple adaptation of the proof of Proposition 2.5.) Since $M_h \equiv M_h^\circ \oplus M_h^\Gamma$, we only need to take care of uniqueness of solution. To this end, let $(\mathbf{q}_h, u_h, \widehat{u}_h) \in \mathbf{V}_h \times W_h \times M_h$ be a solution of

$$\begin{aligned}
(\kappa^{-1}\mathbf{q}_h, \mathbf{r})_{\mathscr{T}_h} - (u_h, \operatorname{div}\mathbf{r})_{\mathscr{T}_h} + \langle \widehat{u}_h, \mathbf{r}\cdot\mathbf{n}\rangle_{\partial\mathscr{T}_h} &= 0 && \forall \mathbf{r} \in \mathbf{V}_h, && (2.77a)\\
(\operatorname{div}\mathbf{q}_h, w)_{\mathscr{T}_h} &= 0 && \forall w \in W_h, && (2.77b)\\
\langle \mathbf{q}_h\cdot\mathbf{n}, \mu\rangle_{\partial\mathscr{T}_h\setminus\Gamma} &= 0 && \forall \mu \in M_h^\circ, && (2.77c)\\
\langle \widehat{u}_h, \mu\rangle_\Gamma &= 0 && \forall \mu \in M_h^\Gamma. && (2.77d)
\end{aligned}$$

Testing these equations with $(\mathbf{q}_h, u_h, -\widehat{u}_h, -\mathbf{q}_h\cdot\mathbf{n})$ and adding the results, we show that $(\kappa^{-1}\mathbf{q}_h, \mathbf{q}_h)_{\mathscr{T}_h} = 0$ and hence $\mathbf{q}_h = \mathbf{0}$. Let us now go back to (2.77a), which after integration by parts and localization on a single element yields for all $K \in \mathscr{T}_h$

$$(\nabla u_h, \mathbf{r})_K = \langle u_h - \widehat{u}_h, \mathbf{r}\cdot\mathbf{n}\rangle_{\partial K} \qquad \forall \mathbf{r} \in \boldsymbol{\mathscr{P}}_k(K). \tag{2.78}$$

We now construct $\mathbf{p} \in \boldsymbol{\mathscr{P}}_k(K)$ satisfying (see (2.72) and Proposition 2.10):

$$\begin{aligned}
(\mathbf{p}, \mathbf{r})_K &= 0 && \forall \mathbf{r} \in \mathscr{N}_{k-2}(K), && (2.79a)\\
\langle \mathbf{p}\cdot\mathbf{n}, \mu\rangle_{\partial K} &= \langle u_h - \widehat{u}_h, \mu\rangle_{\partial K} && \forall \mu \in \mathscr{R}_k(\partial K). && (2.79b)
\end{aligned}$$

Using this function as the test function in (2.78), we prove that

$$0 = (\nabla u_h, \mathbf{p})_K = \langle u_h - \widehat{u}_h, \mathbf{p}\cdot\mathbf{n}\rangle_{\partial K} = \langle u_h - \widehat{u}_h, u_h - \widehat{u}_h\rangle_{\partial K}, \tag{2.80}$$

where we have used $\mu = u_h - \widehat{u}_h \in \mathscr{R}_k(\partial K)$ in (2.79b) and that $\nabla u_h \in \nabla \mathscr{P}_{k-1}(K) \subset \boldsymbol{\mathscr{P}}_{k-2}(K) \subset \mathscr{N}_{k-2}(K)$.

(From here on, everything is just a line-by-line copy of the end of the proof of Proposition 2.5(a), that is, uniqueness of solution of the RT equations.) The equality (2.80) implies that $u_h - \widehat{u}_h = 0$ on ∂K and (2.78) shows then (take $\mathbf{r} = \nabla u_h$) that

$u_h \equiv c_K$ in K and $\widehat{u}_h = u_h \equiv c_K$ on ∂K. Since each interior face value of $\widehat{u}_h$ is reached from different elements, it is easy to see that we have proved that $u_h \equiv c$ and $\widehat{u}_h \equiv c$. However, Eq. (2.77d) implies that $\widehat{u}_h = 0$ on Γ, and the proof of uniqueness of solution of (2.76) is thus finished.

2.4.4 Error Analysis

Energy estimates. We start by redefining the local projections: we take $\boldsymbol{\Pi}\mathbf{q}$ to be the local BDM projection, Πu to be the projection on W_h ($\Pi u|_K := \Pi_{k-1} u$) and Pu to be (again) the orthogonal projection onto M_h. The discrete errors are the same quantities that we defined in (2.30)

$$\boldsymbol{\varepsilon}_h^q := \boldsymbol{\Pi}\mathbf{q} - \mathbf{q}_h \in \mathbf{V}_h, \qquad \varepsilon_h^u := \Pi u - u_h \in W_h, \qquad \widehat{\varepsilon}_h^u := \mathrm{P}u - \widehat{u}_h \in M_h,$$

and the **error equations** differ from those in (2.31)

$$\begin{aligned}
(\kappa^{-1}\boldsymbol{\varepsilon}_h^q, \mathbf{r})_{\mathcal{T}_h} - (\varepsilon_h^u, \operatorname{div}\mathbf{r})_{\mathcal{T}_h} + \langle\widehat{\varepsilon}_h^u, \mathbf{r}\cdot\mathbf{n}\rangle_{\partial\mathcal{T}_h} &= (\kappa^{-1}(\boldsymbol{\Pi}\mathbf{q} - \mathbf{q}), \mathbf{r})_{\mathcal{T}_h} && \forall \mathbf{r} \in \mathbf{V}_h,\\
(\operatorname{div}\boldsymbol{\varepsilon}_h^q, w)_{\mathcal{T}_h} &= 0 && \forall w \in W_h,\\
\langle\boldsymbol{\varepsilon}_h^q\cdot\mathbf{n}, \mu\rangle_{\partial\mathcal{T}_h\setminus\Gamma} &= 0 && \forall \mu \in M_h^\circ,\\
\langle\widehat{\varepsilon}_h^u, \mu\rangle_{\Gamma} &= 0 && \forall \mu \in M_h^\Gamma,
\end{aligned}$$

only in the fact that the spaces and projections have been redefined. The energy estimate (2.33)

$$\|\boldsymbol{\Pi}\mathbf{q} - \mathbf{q}_h\|_{\kappa^{-1}} = \|\boldsymbol{\varepsilon}_h^q\|_{\kappa^{-1}} \leqslant \|\boldsymbol{\Pi}\mathbf{q} - \mathbf{q}\|_{\kappa^{-1}} \tag{2.81}$$

is proved in exactly the same way.

A note on the energy estimate. For purely diffusive problems, the estimate (2.81), together with the approximation properties of the BDM projection (especifically Proposition 2.9(b)) yields optimal convergence $\|\mathbf{q} - \mathbf{q}_h\|_\Omega \lesssim h^{k+1}$. However, for problems with a reaction term

$$\operatorname{div}\mathbf{q} + c\,u = f,$$

the same comments we made in Sect. 2.3.5 still hold, and we can prove

$$\|\boldsymbol{\varepsilon}_h^q\|_{\kappa^{-1}}^2 + |\varepsilon_h^u|_c^2 \leqslant \|\boldsymbol{\Pi}\mathbf{q} - \mathbf{q}\|_{\kappa^{-1}}^2 + |\Pi u - u|_c^2. \tag{2.82}$$

This estimate now implies that $\|\mathbf{q} - \mathbf{q}_h\|_\Omega \lesssim h^k$, due to the influence of the lower order polynomial degree of the space W_h.

More estimates. Using

$$\operatorname{div}\boldsymbol{\xi} = \varepsilon_h^u, \qquad \|\boldsymbol{\xi}\|_{1,\Omega} \leqslant C\|\varepsilon_h^u\|_\Omega,$$

we can repeat the arguments of Sect. 2.2.3 (with the BDM projection now, and taking advantage again of the commutativity property), to reprove (2.37)

$$\|\Pi u - u_h\|_\Omega = \|\varepsilon_h^u\|_\Omega \lesssim \|\mathbf{q} - \mathbf{q}_h\|_{\kappa^{-1}}. \tag{2.83}$$

Taking

$$\mathbf{r} \in \mathscr{P}_k(K), \qquad \mathbf{r} \cdot \mathbf{n} = \widehat{\varepsilon}_h^u, \qquad \|\mathbf{r}\|_K \lesssim h_K^{1/2} \|\widehat{\varepsilon}_h^u\|_{\partial K},$$

we can also prove (2.40) for BDM

$$\|\mathrm{P}u - \widehat{u}_h\|_h = \|\widehat{\varepsilon}_h^u\|_h \lesssim \|\varepsilon_h^u\|_\Omega + h\|\mathbf{q} - \mathbf{q}_h\|_{\kappa^{-1}}. \tag{2.84}$$

Duality estimates. Once again, we consider the dual problem

$$\begin{aligned} \kappa^{-1}\boldsymbol{\xi} - \nabla\theta &= \mathbf{0} && \text{in } \Omega, \\ \operatorname{div} \boldsymbol{\xi} &= \varepsilon_h^u && \text{in } \Omega, \\ \theta &= 0 && \text{on } \Gamma, \end{aligned}$$

and assume the following **regularity hypothesis**: there exists C such that

$$\|\boldsymbol{\xi}\|_{1,\Omega} + \|\theta\|_{2,\Omega} \leqslant C\|\varepsilon_h^u\|_\Omega.$$

The arguments of Sect. 2.2.4 can be repeated and

$$\|\varepsilon_h^u\|_\Omega \lesssim h\,(\|\mathbf{q}_h - \mathbf{q}\|_{\kappa^{-1}} + \|f - \Pi f\|_\Omega)$$

follows with the same proof. For $k \geqslant 2$, we can do slightly better when we bound

$$|(f - \Pi f, \theta - \Pi\theta)_\Omega| \lesssim h^2\|f - \Pi f\|_\Omega |\theta|_{2,\Omega} \lesssim h^2\|f - \Pi f\|_\Omega \|\varepsilon_h^u\|_\Omega.$$

(This estimate does not work for $k = 1$, since then Π is the projection on piecewise constants and cannot deliver the h^2 term.) The general case can then be presented as

$$\|\varepsilon_h^u\|_\Omega \lesssim h\,\|\mathbf{q}_h - \mathbf{q}\|_{\kappa^{-1}} + h^{\min\{k,2\}}\|f - \Pi f\|_\Omega. \tag{2.85}$$

Optimal convergence. Approximation of the projections used in the projection-based analysis can be summarized (for smoothest solutions) as

$$\|\boldsymbol{\Pi}\mathbf{q} - \mathbf{q}\|_\Omega \lesssim h^{k+1}, \qquad \|\Pi u - u\|_\Omega \lesssim h^k, \qquad \|\mathrm{P}u - u\|_h \lesssim h^{k+1}.$$

With everything in our favor, the BDM Eqs. (2.72) provide the following error estimates:

$$
\begin{aligned}
\|\mathbf{q}-\mathbf{q}_h\|_\Omega &\lesssim h^{k+1}, & \|\boldsymbol{\Pi}\mathbf{q}-\mathbf{q}_h\|_\Omega &\lesssim h^{k+1}, && \text{(see (2.81))}\\
\|u-u_h\|_\Omega &\lesssim h^{k}, & \|\Pi u-u_h\|_\Omega &\lesssim h^{k+\min\{k,2\}}, && \text{(see (2.85))}\\
\|u-\widehat{u}_h\|_h &\lesssim h^{k+1}, & \|\mathrm{P}u-\widehat{u}_h\|_h &\lesssim h^{k+\min\{k,2\}}, && \text{(see (2.85)\&(2.84))}\\
\|\mathbf{q}\cdot\mathbf{n}-\mathbf{q}_h\cdot\mathbf{n}\|_h &\lesssim h^{k+1}, & \|\boldsymbol{\Pi}\mathbf{q}\cdot\mathbf{n}-\mathbf{q}_h\cdot\mathbf{n}\|_h &\lesssim h^{k+1}.
\end{aligned}
$$

Let it be noted that when there is a reaction term in the equation, convergence for $\mathbf{q}_h$ is subject to the estimate (2.82) and reduced to h^k. This bound is dragged to all the superconvergence estimates.

Chapter 3
The Hybridizable Discontinuous Galerkin Method

In this section, we show how the spaces of RT and BDM can be balanced to have an equal polynomial degree. Stability will be restored using a discrete stabilization (not penalization) function. This is how local quantities of RT, BDM, and HDG methods compare. Note that there is no natural finite element structure for $\mathbf{q}_h$, where we can recognize boundary and internal degrees of freedom. Instead, we will have a projection that integrates $(\mathbf{q}_h, u_h)$ into the same structure.

degree	$\mathbf{q}_h$	u_h	bd. d.o.f.	int. d.o.f.
$k \geqslant 0$	$\mathcal{RT}_k(K) = \boldsymbol{\mathcal{P}}_k(K) + \mathbf{m}\widetilde{\mathcal{P}}_k(K)$	$\mathcal{P}_k(K)$	$\mathcal{R}_k(\partial K)$	$\boldsymbol{\mathcal{P}}_{k-1}(K)$
$k \geqslant 1$	$\boldsymbol{\mathcal{P}}_k(K)$	$\mathcal{P}_{k-1}(K)$	$\mathcal{R}_k(\partial K)$	$\mathcal{N}_{k-2}(K)$
$k \geqslant 0$	$\boldsymbol{\mathcal{P}}_k(K)$	$\mathcal{P}_k(K)$	N.A.	N.A.

Let us start with some small talk. The Hybridizable Discontinuous Galerkin method can be understood as a further development of the Local Discontinuous Galerkin method, one of the many DG schemes covered in the celebrated *framework-style* paper of Arnold et al. [2]. While the trail of papers is not entirely obvious, premonitions of what was about to happen can be found in the treatment of hybridized mixed methods by Cockburn and Gopalakrishnan [41]. Some time later, this fructified in another long framework-style paper of the previous authors and Lazarov [43], setting the bases for a full development of HDG methods. Cockburn and collaborators have been pushing the limits of applicability of HDG ideas to many problems in continuum mechanics and physics. The analysis, as will be shown here, is based on a particular definition of a projection tailored to the HDG equations: its first occurrence was due to Cockburn et al. in [45].

S. Du and F.-J. Sayas, *An Invitation to the Theory of the Hybridizable Discontinuous Galerkin Method*, SpringerBriefs in Mathematics,
https://doi.org/10.1007/978-3-030-27230-2_3

3.1 The HDG Method

For $k \geqslant 0$, consider the spaces

$$\mathbf{V}_h := \prod_{K\in\mathscr{T}_h} \mathscr{P}_k(K), \qquad W_h := \prod_{K\in\mathscr{T}_h} \mathscr{P}_k(K), \qquad M_h := \prod_{e\in\mathscr{E}_h} \mathscr{P}_k(e),$$

and the subspace decomposition $M_h = M_h^\circ \oplus M_h^\Gamma$. Consider also the **stabilization function**

$$\tau \in \prod_{K\in\mathscr{T}_h} \mathscr{R}_0(\partial K), \qquad \tau \geqslant 0 \qquad \tau|_{\partial K} \neq 0 \ \ \forall K.$$

We look for

$$(\mathbf{q}_h, u_h, \widehat{u}_h) \in \mathbf{V}_h \times W_h \times M_h, \tag{3.1a}$$

satisfying

$$\begin{aligned}
&(\kappa^{-1}\mathbf{q}_h, \mathbf{r})_{\mathscr{T}_h} - (u_h, \operatorname{div}\mathbf{r})_{\mathscr{T}_h} + \langle \widehat{u}_h, \mathbf{r}\cdot\mathbf{n}\rangle_{\partial\mathscr{T}_h} = 0 && \forall \mathbf{r} \in \mathbf{V}_h, && (3.1b)\\
&(\operatorname{div}\mathbf{q}_h, w)_{\mathscr{T}_h} + \langle \tau(u_h - \widehat{u}_h), w\rangle_{\partial\mathscr{T}_h} = (f, w)_{\mathscr{T}_h} && \forall w \in W_h, && (3.1c)\\
&\langle \mathbf{q}_h\cdot\mathbf{n} + \tau(u_h - \widehat{u}_h), \mu\rangle_{\partial\mathscr{T}_h\setminus\Gamma} = 0 && \forall \mu \in M_h^\circ, && (3.1d)\\
&\langle \widehat{u}_h, \mu\rangle_\Gamma = \langle g, \mu\rangle_\Gamma && \forall \mu \in M_h^\Gamma. && (3.1e)
\end{aligned}$$

Some comments. Equations (3.1b) and (3.1c) are local, given the fact that the spaces are discontinuous. The first of them is the same equation (with different spaces) as in the RT and BDM methods. If τ were to be zero (this is not allowed in our choice of spaces), Eq. (3.1c) would be the same equation that we had in RT and BDM. Note that, after integration by parts, we can also write (3.1c) as

$$-(\mathbf{q}_h, \nabla w)_{\mathscr{T}_h} + \langle \mathbf{q}_h\cdot\mathbf{n} + \tau(u_h - \widehat{u}_h), w\rangle_{\partial\mathscr{T}_h} = (f, w)_{\mathscr{T}_h} \qquad \forall w \in W_h,$$

where the numerical flux

$$\widehat{\mathbf{q}}_h\cdot\mathbf{n} := \mathbf{q}_h\cdot\mathbf{n} + \tau(u_h - \widehat{u}_h) \in \mathscr{R}_k(\partial K) \qquad \forall K, \tag{3.2}$$

makes an appearance. Equation (3.1d) imposes that this numerical flux is "single-valued" on all internal faces (actually, normal components cancel each other), so that the numerical flux $\widehat{\mathbf{q}}_h\cdot\mathbf{n}$ can be identified with an element of M_h.

Proposition 3.1 *Equations* (3.1) *have a unique solution.*

Proof (What follows is a slight adaptation of the proofs of Propositions 2.5(a) (RT) and 2.11 (BDM) to the HDG equations.) We only need to show that any solution $(\mathbf{q}_h, u_h, \widehat{u}_h) \in \mathbf{V}_h \times W_h \times M_h$ of the homogeneous equations

$$
\begin{aligned}
(\kappa^{-1}\mathbf{q}_h, \mathbf{r})_{\mathscr{T}_h} - (u_h, \operatorname{div}\mathbf{r})_{\mathscr{T}_h} + \langle \widehat{u}_h, \mathbf{r}\cdot\mathbf{n}\rangle_{\partial\mathscr{T}_h} &= 0 \quad \forall \mathbf{r} \in \mathbf{V}_h,\\
(\operatorname{div}\mathbf{q}_h, w)_{\mathscr{T}_h} + \langle \tau(u_h - \widehat{u}_h), w\rangle_{\partial\mathscr{T}_h} &= 0 \quad \forall w \in W_h,\\
\langle \mathbf{q}_h\cdot\mathbf{n} + \tau(u_h - \widehat{u}_h), \mu\rangle_{\partial\mathscr{T}_h\setminus\Gamma} &= 0 \quad \forall \mu \in M_h^\circ,\\
\langle \widehat{u}_h, \mu\rangle_\Gamma &= 0 \quad \forall \mu \in M_h^\Gamma,
\end{aligned}
$$

vanishes. Testing these equations with $(\mathbf{q}_h, u_h, -\widehat{u}_h, -\mathbf{q}_h\cdot\mathbf{n} - \tau(u_h - \widehat{u}_h))$ and adding the results, we easily prove that

$$
(\kappa^{-1}\mathbf{q}_h, \mathbf{q}_h)_{\mathscr{T}_h} + \langle \tau(u_h - \widehat{u}_h), u_h - \widehat{u}_h\rangle_{\partial\mathscr{T}_h} = 0,
$$

and therefore $\mathbf{q}_h = \mathbf{0}$, $\tau(u_h - \widehat{u}_h) = 0$ (we have used that $\tau \geqslant 0$) and

$$
(\nabla u_h, \mathbf{r})_K + \langle u_h - \widehat{u}_h, \mathbf{r}\cdot\mathbf{n}\rangle_{\partial K} = 0 \quad \forall \mathbf{r} \in \mathscr{P}_k(K) \quad \forall K. \tag{3.3}
$$

In particular, we have

$$
\langle u_h - \widehat{u}_h, \mathbf{r}\cdot\mathbf{n}\rangle_{\partial K} = 0 \quad \forall \mathbf{r} \in \mathscr{P}_k^\perp(K).
$$

This implies (Lemma 2.2) that $u_h - \widehat{u}_h = v$ on ∂K, where $v \in \mathscr{P}_k^\perp(K)$. However, since $\tau(u_h - \widehat{u}_h) = 0$ and τ is at least positive in one face of K, then (by Lemma 2.1(a)) necessarily $v = 0$ and thus $u_h - \widehat{u}_h = 0$ on ∂K. Testing then (3.3) with $\mathbf{r} = \nabla u_h$, we show that $u_h \equiv c_K$ on K and $u_h = \widehat{u}_h \equiv c_K$ on ∂K. Proceeding as in the proof of Proposition 2.5, we show that $u_h = 0$ and $\widehat{u}_h = 0$.

3.2 The HDG Projection

The analysis of the HDG method will follow the same pattern we have employed in the analysis of RT and BDM. We start by defining a tailored projection onto the discrete spaces that will be used to write error equations that mimic those of the hybridizable mixed methods. As opposed to the two separate projections for $\mathbf{q}$ and u that were used in RT and BDM, here the projection will be defined for the pair $(\mathbf{q}, u)$. However, we will still denote $(\Pi^{\mathrm{HDG}}\mathbf{q}, \Pi^{\mathrm{HDG}}u)$, as if these projections were defined separately: correct, but cumbersome, notation would express these elements as components of a single operator.

The HDG projection. Given sufficiently smooth $(\mathbf{q}, u) : K \to \mathbb{R}^d \times \mathbb{R}$, we define

$$
(\boldsymbol{\Pi}^{\mathrm{HDG}}\mathbf{q}, \Pi^{\mathrm{HDG}}u) := (\boldsymbol{\Pi}_q^{\mathrm{HDG}}(\mathbf{q}, u), \Pi_u^{\mathrm{HDG}}(\mathbf{q}, u)) \in \mathscr{P}_k(K) \times \mathscr{P}_k(K)
$$

as the solution to the equations

$$(\boldsymbol{\Pi}^{\mathrm{HDG}}\mathbf{q}, \mathbf{r})_K = (\mathbf{q}, \mathbf{r})_K \qquad \forall \mathbf{r} \in \boldsymbol{\mathscr{P}}_{k-1}(K), \quad (3.4a)$$
$$(\Pi^{\mathrm{HDG}}u, v)_K = (u, v)_K \qquad \forall v \in \mathscr{P}_{k-1}(K), \quad (3.4b)$$
$$\langle \boldsymbol{\Pi}^{\mathrm{HDG}}\mathbf{q}\cdot\mathbf{n} + \tau\Pi^{\mathrm{HDG}}u, \mu\rangle_{\partial K} = \langle \mathbf{q}\cdot\mathbf{n} + \tau\, u, \mu\rangle_{\partial K} \qquad \forall \mu \in \mathscr{R}_k(\partial K). \quad (3.4c)$$

Proposition 3.2 (Definition of the HDG projection) *Equations* (3.4) *are uniquely solvable.*

Proof We first remark that

$$\dim \boldsymbol{\mathscr{P}}_k(K) + \dim \mathscr{P}_k(K) = \dim \boldsymbol{\mathscr{P}}_{k-1}(K) + \dim \mathscr{P}_{k-1}(K) + \dim \mathscr{R}_k(\partial K),$$

and therefore we only need to show uniqueness. Let then $(\mathbf{q}, u) \in \boldsymbol{\mathscr{P}}_k(K) \times \mathscr{P}_k(K)$ satisfy

$$(\mathbf{q}, \mathbf{r})_K = 0 \qquad \forall \mathbf{r} \in \boldsymbol{\mathscr{P}}_{k-1}(K), \quad (3.5a)$$
$$(u, v)_K = 0 \qquad \forall v \in \mathscr{P}_{k-1}(K), \quad (3.5b)$$
$$\langle \mathbf{q}\cdot\mathbf{n} + \tau u, \mu\rangle_{\partial K} = 0 \qquad \forall \mu \in \mathscr{R}_k(\partial K). \quad (3.5c)$$

Then $\mathbf{q} \in \boldsymbol{\mathscr{P}}_k^{\perp}(K)$ and $u \in \mathscr{P}_k^{\perp}(K)$. Testing (3.5c) with $u|_{\partial K}$ and using Lemma 2.2, we prove that

$$0 = \langle \mathbf{q}\cdot\mathbf{n} + \tau\, u, u\rangle_{\partial K} = \langle \tau\, u, u\rangle_{\partial K} = \langle \tau^{1/2}u, \tau^{1/2}u\rangle_{\partial K},$$

and therefore $\tau^{1/2}u = 0$ (we have used here that $\tau \geqslant 0$). We can now test (3.5c) with $\mathbf{q}\cdot\mathbf{n}$ to prove that $\mathbf{q}\cdot\mathbf{n} = 0$ on ∂K. By Lemma 2.1(b), it follows that $\mathbf{q} = \mathbf{0}$. On the other hand, $\tau u = 0$ on ∂K and we have assumed that $\tau > 0$ in at least one face of K. Lemma 2.1(a) proves then that $u = 0$.

Weak commutativity. For general $(\mathbf{q}, u)$ and $v \in \mathscr{P}_k(K)$,

$$\begin{aligned}(\operatorname{div}\boldsymbol{\Pi}^{\mathrm{HDG}}\mathbf{q}, v)_K &= \langle \boldsymbol{\Pi}^{\mathrm{HDG}}\mathbf{q}\cdot\mathbf{n}, v\rangle_{\partial K} - (\boldsymbol{\Pi}^{\mathrm{HDG}}\mathbf{q}, \nabla v)_K \\ &= \langle \mathbf{q}\cdot\mathbf{n} - \tau(\Pi^{\mathrm{HDG}}u - u), v\rangle_{\partial K} - (\mathbf{q}, \nabla v)_K \qquad \text{(by (3.4))} \\ &= (\operatorname{div}\mathbf{q}, v)_K - \langle \tau(\Pi^{\mathrm{HDG}}u - u), v\rangle_{\partial K},\end{aligned}$$

which can be rewritten as

$$(\operatorname{div}\boldsymbol{\Pi}^{\mathrm{HDG}}\mathbf{q}, v)_K + \langle \tau\, \Pi^{\mathrm{HDG}}u, v\rangle_{\partial K} = (\operatorname{div}\mathbf{q}, v)_K + \langle \tau\, u, v\rangle_{\partial K} \quad (3.6)$$

for all $v \in \mathscr{P}_k(K)$. Compare this result with the clean commutativity properties of the RT (2.8) and BDM (2.74) projections.

Change to the reference element. Let $\check{\tau} := |a_K|\tau \circ \mathrm{F}_K|_{\partial\widehat{K}}$. Consider then the projection $(\widehat{\boldsymbol{\Pi}}^{\mathrm{HDG}}, \widehat{\Pi}^{\mathrm{HDG}})$ associated to the stabilization function $\check{\tau}$. It is then easy to show that

$$(\widehat{\boldsymbol{\Pi}^{\mathrm{HDG}}\mathbf{q}}, \widehat{\Pi^{\mathrm{HDG}}u}) = (\widehat{\boldsymbol{\Pi}}^{\mathrm{HDG}}\widehat{\mathbf{q}}, \widehat{\Pi}^{\mathrm{HDG}}\widehat{u}). \tag{3.7}$$

Decoupling of the equations. Let us first use $\mu = w|_{\partial K}$ in (3.4c), where $w \in \mathscr{P}_k^{\perp}(K)$. We then have

$$\begin{aligned}
\langle \tau(\Pi^{\mathrm{HDG}}u - u), w\rangle_{\partial K} &= \langle(\mathbf{q} - \boldsymbol{\Pi}^{\mathrm{HDG}}\mathbf{q}) \cdot \mathbf{n}, w\rangle_{\partial K} \\
&= (\operatorname{div}\mathbf{q} - \operatorname{div}\boldsymbol{\Pi}^{\mathrm{HDG}}\mathbf{q}, w)_K + (\mathbf{q} - \boldsymbol{\Pi}^{\mathrm{HDG}}\mathbf{q}, \nabla w)_K \\
&= (\operatorname{div}\mathbf{q}, w)_K. \qquad \text{(by (3.4a) and since } w \in \mathscr{P}_k^{\perp}(K))
\end{aligned}$$

Therefore, the solution of (3.4) also satisfies

$$\begin{aligned}
(\Pi^{\mathrm{HDG}}u, w)_K &= (u, w)_K & \forall w \in \mathscr{P}_{k-1}(K), \quad (3.8a)\\
\langle \tau\Pi^{\mathrm{HDG}}u, w\rangle_{\partial K} &= \langle \tau u, w\rangle_{\partial K} + (\operatorname{div}\mathbf{q}, w)_K & \forall w \in \mathscr{P}_k^{\perp}(K), \quad (3.8b)
\end{aligned}$$

and

$$\begin{aligned}
(\boldsymbol{\Pi}^{\mathrm{HDG}}\mathbf{q}, \mathbf{r})_K &= (\mathbf{q}, \mathbf{r})_K & \forall \mathbf{r} \in \boldsymbol{\mathscr{P}}_{k-1}(K), \quad (3.8c)\\
\langle \boldsymbol{\Pi}^{\mathrm{HDG}}\mathbf{q} \cdot \mathbf{n}, \mu\rangle_{\partial K\setminus e} &= \langle \mathbf{q} \cdot \mathbf{n}, \mu\rangle_{\partial K\setminus e} & \\
&\quad + \langle \tau(u - \Pi^{\mathrm{HDG}}u), \mu\rangle_{\partial K\setminus e} & \forall \mu \in \mathscr{R}_k(\partial K\setminus e), \quad (3.8d)
\end{aligned}$$

where e is any face of ∂K. Note that Eqs. (3.8a)–(3.8b) are uniquely solvable by Lemma 2.1(a) and we show that $\Pi^{\mathrm{HDG}}u$ depends on u and $\operatorname{div}\mathbf{q}$. Equations (3.8c) and (3.8d) are also uniquely solvable, as follows from a comment at the end of the proof of Lemma 2.1(b), namely, if $\mathbf{q} \in \boldsymbol{\mathscr{P}}_k^{\perp}(K)$ and $\mathbf{q} \cdot \mathbf{n} = 0$ on $\partial K\setminus e$ (e is any face of ∂K), then $\mathbf{q} = \mathbf{0}$.

The single-face HDG method. A particular choice of the stabilization function τ was given in [30]. It consists of choosing one particular $e_K \in \mathscr{E}(K)$ and taking $\tau_{\partial K} > 0$ in e_K and $\tau_K \equiv 0$ in $\partial K\setminus e$. Because of this particular construction of the stabilization function, the method is called the single-face HDG. This shows (take $e = e_K$ in (3.8d)) that the vector part of the HDG projection is completely decoupled from the scalar part for the SF–HDG case, and that it does not depend on τ.

3.3 Estimates for the HDG Projection

Notation. For some forthcoming arguments, it will be useful to isolate the face

$$\widehat{e} \in \mathscr{E}(\widehat{K}), \qquad \widehat{e} \subset \{\mathbf{x} \in \mathbb{R}^d \,:\, \mathbf{x} \cdot (1, \ldots, 1) = 1\}.$$

From this moment on, *the symbol $\lesssim$ will include independence of the parameters τ as well.*

Proposition 3.3 (Estimate on the reference domain—Part I) *Given* u, f, *and* $0 \neq \breve{\tau} \in \mathscr{R}_0(\partial \widehat{K})$, *such that* $\breve{\tau} \geqslant 0$ *and* $\breve{\tau}_\circ := \breve{\tau}|_{\widehat{e}} > 0$, *we define* $\widehat{\Pi} u \in \mathscr{P}_k(\widehat{K})$ *by solving the equations*

$$(\widehat{\Pi} u, w)_{\widehat{K}} = (u, w)_{\widehat{K}} \qquad \forall w \in \mathscr{P}_{k-1}(\widehat{K}), \tag{3.9a}$$

$$\langle \breve{\tau} \widehat{\Pi} u, w\rangle_{\partial \widehat{K}} = \langle \breve{\tau} u, w\rangle_{\partial \widehat{K}} + (f, w)_{\widehat{K}} \qquad \forall w \in \mathscr{P}_k^\perp(\widehat{K}). \tag{3.9b}$$

Then

$$\|\widehat{\Pi} u\|_{\widehat{K}} \lesssim \breve{\tau}_\circ^{-1}\big(\|\breve{\tau}\|_{L^\infty} \|u\|_{1,\widehat{K}} + \|f\|_{\widehat{K}}\big), \tag{3.10}$$

$$\|u - \widehat{\Pi} u\|_{\widehat{K}} \lesssim \breve{\tau}_\circ^{-1}\big(\|\breve{\tau}\|_{L^\infty} |u|_{k+1,\widehat{K}} + |f|_{k,\widehat{K}}\big). \tag{3.11}$$

Proof Let $\delta := \widehat{\Pi} u - \widehat{\Pi}_k u$ ($\widehat{\Pi}_k$ is the L^2 projection onto $\mathscr{P}_k(\widehat{K})$), and note that $\delta \in \mathscr{P}_k^\perp(\widehat{K})$ by (3.9a). Then

$$\begin{aligned}
\|\delta\|_{\widehat{K}}^2 &\lesssim \|\delta\|_{\widehat{e}}^2 && \text{(conseq of Lemma 2.1(a))}\\
&= \breve{\tau}_\circ^{-1} \langle \breve{\tau} \delta, \delta\rangle_{\widehat{e}} && (\breve{\tau}|_{\widehat{e}} = \breve{\tau}_\circ)\\
&\leqslant \breve{\tau}_\circ^{-1} \langle \breve{\tau} \delta, \delta\rangle_{\partial \widehat{K}} && (\breve{\tau} \geqslant 0)\\
&= \breve{\tau}_\circ^{-1} \big(\langle \breve{\tau}(u - \widehat{\Pi}_k u), \delta\rangle_{\partial \widehat{K}} + (f, \delta)_{\widehat{K}}\big) && (\delta \in \mathscr{P}_k^\perp \text{ and (3.9b)))}\\
&\leqslant \breve{\tau}_\circ^{-1} \big(\|\breve{\tau}\|_{L^\infty} \|u - \widehat{\Pi}_k u\|_{\partial \widehat{K}} \|\delta\|_{\partial \widehat{K}} + \|f\|_{\widehat{K}} \|\delta\|_{\widehat{K}}\big)\\
&\lesssim \breve{\tau}_\circ^{-1} \big(\|\breve{\tau}\|_{L^\infty} \|u - \widehat{\Pi}_k u\|_{\partial \widehat{K}} + \|f\|_{\widehat{K}}\big) \|\delta\|_{\widehat{K}} && \text{(finite dimensions)}\\
&\lesssim \breve{\tau}_\circ^{-1} \big(\|\breve{\tau}\|_{L^\infty} \|u - \widehat{\Pi}_k u\|_{1,\widehat{K}} + \|f\|_{\widehat{K}}\big) \|\delta\|_{\widehat{K}} && \text{(trace theorem)}\\
&\lesssim \breve{\tau}_\circ^{-1} \big(\|\breve{\tau}\|_{L^\infty} \|u\|_{1,\widehat{K}} + \|f\|_{\widehat{K}}\big) \|\delta\|_{\widehat{K}}. && \text{(finite dimensions)}
\end{aligned}$$

Therefore

$$\|\widehat{\Pi} u\|_{\widehat{K}} \leqslant \|\widehat{\Pi}_k u\|_{\widehat{K}} + \|\delta\|_{\widehat{K}} \lesssim \|u\|_{\widehat{K}} + \breve{\tau}_\circ^{-1} \big(\|\breve{\tau}\|_{L^\infty} \|u\|_{1,\widehat{K}} + \|f\|_{\widehat{K}}\big),$$

and (3.10) is thus proved. At the same time, note that we can substitute (3.9b) by

$$\langle \breve{\tau} \widehat{\Pi} u, w\rangle_{\partial \widehat{K}} = \langle \breve{\tau} u, w\rangle_{\partial \widehat{K}} + (f - \widehat{\Pi}_{k-1} f, w)_{\widehat{K}} \qquad \forall w \in \mathscr{P}_k^\perp(\widehat{K}).$$

Returning to our previous argument, we have

$$\|\widehat{\Pi} u - \widehat{\Pi}_k u\|_{\widehat{K}} \lesssim \breve{\tau}_\circ^{-1} \big(\|\breve{\tau}\|_{L^\infty} \|u - \widehat{\Pi}_k u\|_{1,\widehat{K}} + \|f - \widehat{\Pi}_{k-1} f\|_{\widehat{K}}\big),$$

and (3.11) follows from a compactness (Bramble–Hilbert style) argument.

Proposition 3.4 (Estimate on the reference domain—Part II) *Given* ε, $\mathbf{q}$, *and* $0 \neq \breve{\tau} \in \mathscr{R}_0(\partial \widehat{K})$, $\breve{\tau} \geqslant 0$, *we define* $\widehat{\boldsymbol{\Pi}} \mathbf{q} \in \boldsymbol{\mathscr{P}}_k(\widehat{K})$ *by solving the equations*

$$(\widehat{\boldsymbol{\Pi}}\mathbf{q}, \mathbf{r})_{\widehat{K}} = (\mathbf{q}, \mathbf{r})_{\widehat{K}} \qquad \forall \mathbf{r} \in \boldsymbol{\mathscr{P}}_{k-1}(\widehat{K}), \tag{3.12a}$$

$$\langle \widehat{\boldsymbol{\Pi}}\mathbf{q} \cdot \widehat{\mathbf{n}}, \mu \rangle_{\partial\widehat{K}\setminus\widehat{e}} = \langle \mathbf{q} \cdot \widehat{\mathbf{n}} + \check{\tau}\varepsilon, \mu \rangle_{\partial\widehat{K}\setminus\widehat{e}} \qquad \forall \mu \in \mathscr{R}_k(\partial\widehat{K}\setminus\widehat{e}). \tag{3.12b}$$

With this definition, we have the estimates

$$\|\widehat{\boldsymbol{\Pi}}\mathbf{q}\|_{\widehat{K}} \lesssim \|\mathbf{q}\|_{1,\widehat{K}} + \|\check{\tau}\|_{L^\infty(\partial\widehat{K}\setminus\widehat{e})}\|\varepsilon\|_{\partial\widehat{K}}, \tag{3.13}$$

$$\|\mathbf{q} - \widehat{\boldsymbol{\Pi}}\mathbf{q}\|_{\widehat{K}} \lesssim |\mathbf{q}|_{k+1,\widehat{K}} + \|\check{\tau}\|_{L^\infty(\partial\widehat{K}\setminus\widehat{e})}\|\varepsilon\|_{\partial\widehat{K}}. \tag{3.14}$$

Proof The stability estimate (3.13) follows from a simple finite-dimensional argument. To prove (3.14), we compare with the componentwise L^2 projection onto $\boldsymbol{\mathscr{P}}_k(\widehat{K})$. Since $\boldsymbol{\delta} := \widehat{\boldsymbol{\Pi}}\mathbf{q} - \widehat{\boldsymbol{\Pi}}_k\mathbf{q} \in \boldsymbol{\mathscr{P}}_k^\perp(\widehat{K})$, we can use an argument based on Lemma 2.1(b) to bound

$$\begin{aligned}
\|\boldsymbol{\delta}\|_{\widehat{K}}^2 &\lesssim \|\boldsymbol{\delta}\cdot\widehat{\mathbf{n}}\|_{\partial\widehat{K}\setminus\widehat{e}}^2 = \langle \boldsymbol{\delta}\cdot\widehat{\mathbf{n}}, \boldsymbol{\delta}\cdot\widehat{\mathbf{n}}\rangle_{\partial\widehat{K}\setminus\widehat{e}} \\
&= \langle \mathbf{q}\cdot\widehat{\mathbf{n}} - \widehat{\boldsymbol{\Pi}}_k\mathbf{q}\cdot\widehat{\mathbf{n}} + \check{\tau}\varepsilon, \boldsymbol{\delta}\cdot\widehat{\mathbf{n}}\rangle_{\partial\widehat{K}\setminus\widehat{e}} \qquad \text{(by (3.12b))} \\
&\leqslant \left(\|\mathbf{q}\cdot\widehat{\mathbf{n}} - \widehat{\boldsymbol{\Pi}}_k\mathbf{q}\cdot\widehat{\mathbf{n}}\|_{\partial\widehat{K}} + \|\check{\tau}\|_{L^\infty(\partial\widehat{K}\setminus\widehat{e})}\|\varepsilon\|_{\partial\widehat{K}}\right)\|\boldsymbol{\delta}\|_{\partial\widehat{K}} \\
&\lesssim \left(\|\mathbf{q} - \widehat{\boldsymbol{\Pi}}_k\mathbf{q}\|_{1,\widehat{K}} + \|\check{\tau}\|_{L^\infty(\partial\widehat{K}\setminus\widehat{e})}\|\varepsilon\|_{\partial\widehat{K}}\right)\|\boldsymbol{\delta}\|_{\widehat{K}},
\end{aligned}$$

where we used trace theorem and finite-dimensional arguments for the last step. Now the result follows from a compactness argument.

Proposition 3.5 (Estimates for the HDG projection) *Given* $\mathbf{q}, u$, *and* $0 \neq \tau \in \mathscr{R}_0(\partial K)$, $\tau \geqslant 0$,

$$\|u - \Pi^{\mathrm{HDG}}u\|_K \lesssim h_K^{k+1}\left(|u|_{k+1,K} + \tau_{\max}^{-1}|\operatorname{div}\mathbf{q}|_{k,K}\right), \tag{3.15a}$$

$$\|\mathbf{q} - \boldsymbol{\Pi}^{\mathrm{HDG}}\mathbf{q}\|_K \lesssim h_K^{k+1}\left(|\mathbf{q}|_{k+1,K} + \tau^\star|u|_{k+1,K}\right), \tag{3.15b}$$

with $\tau_{\max} := \|\tau\|_{L^\infty}$ *and* $\tau^\star := \|\tau\|_{L^\infty(\partial K\setminus e)}$, *where* $\tau|_e = \tau_{\max}$.

Proof The estimate for u follows from Proposition 3.3 and a scaling argument. Note first that by (3.7) we can study the error on the reference element. Doing as in (3.8), we have

$$\begin{aligned}
(\widehat{\Pi}^{\mathrm{HDG}}\widehat{u}, w)_{\widehat{K}} &= (\widehat{u}, w)_{\widehat{K}} && \forall w \in \mathscr{P}_{k-1}(\widehat{K}), \\
\langle \check{\tau}\widehat{\Pi}^{\mathrm{HDG}}\widehat{u}, w\rangle_{\partial\widehat{K}} &= \langle \check{\tau}\widehat{u}, w\rangle_{\partial\widehat{K}} + (\widehat{\operatorname{div}\mathbf{q}}, w)_{\widehat{K}} && \forall w \in \mathscr{P}_k^\perp(\widehat{K}), \\
(\widehat{\boldsymbol{\Pi}}^{\mathrm{HDG}}\widehat{\mathbf{q}}, \mathbf{r})_{\widehat{K}} &= (\widehat{\mathbf{q}}, \mathbf{r})_{\widehat{K}} && \forall \mathbf{r} \in \boldsymbol{\mathscr{P}}_{k-1}(\widehat{K}), \\
\langle \widehat{\boldsymbol{\Pi}}^{\mathrm{HDG}}\widehat{\mathbf{q}}\cdot\widehat{\mathbf{n}}, \mu\rangle_{\partial\widehat{K}\setminus\widehat{e}} &= \langle \widehat{\mathbf{q}}\cdot\widehat{\mathbf{n}}, \mu\rangle_{\partial\widehat{K}\setminus\widehat{e}} \\
&\quad + \langle \check{\tau}(\widehat{u} - \widehat{\Pi}^{\mathrm{HDG}}\widehat{u}), \mu\rangle_{\partial\widehat{K}\setminus\widehat{e}} && \forall \mu \in \mathscr{R}_k(\partial\widehat{K}\setminus\widehat{e}),
\end{aligned}$$

where the transformation $\mathrm{F}_K : \widehat{K} \to K$ is chosen so that $\mathrm{F}_K(\widehat{e}) = e$, where $e \in \mathscr{E}(K)$ is such that $\tau|_e = \tau_{\max}$. We first apply Proposition 3.3 with

$$f = \widetilde{\operatorname{div} \mathbf{q}} = \widehat{\operatorname{div}}\, \widehat{\mathbf{q}} = |J_K| \widehat{\operatorname{div} \mathbf{q}}, \tag{3.16a}$$

$$\breve{\tau}_\circ \approx \tau_{\max} h_K^{d-1} \approx \|\breve{\tau}\|_{L^\infty}, \quad \|\breve{\tau}\|_{L^\infty(\partial \widehat{K} \setminus \widehat{e})} \lesssim \tau^\star h_K^{d-1}. \tag{3.16b}$$

It follows that

$$\begin{aligned}
&\|u - \Pi^{\mathrm{HDG}} u\|_K \\
&\quad \approx h_K^{\frac{d}{2}} \|\widehat{u} - \widehat{\Pi}^{\mathrm{HDG}} \widehat{u}\|_{\widehat{K}} && \text{(by (1.7) and (3.7))} \\
&\quad \lesssim h_K^{\frac{d}{2}} \breve{\tau}_\circ^{-1} \big(\|\breve{\tau}\|_{L^\infty} |\widehat{u}|_{k+1,\widehat{K}} + |\widehat{\operatorname{div}}\, \widehat{\mathbf{q}}|_{k,\widehat{K}} \big) && \text{(by Proposition 3.3)} \\
&\quad \lesssim h_K^{\frac{d}{2}} |\widehat{u}|_{k+1,\widehat{K}} + \tau_{\max}^{-1} h_K^{1-\frac{d}{2}} |J_K| |\widehat{\operatorname{div} \mathbf{q}}|_{k,\widehat{K}} && \text{(by (3.16))} \\
&\quad \lesssim h_K^{\frac{d}{2}} |\widehat{u}|_{k+1,\widehat{K}} + \tau_{\max}^{-1} h_K^{1+\frac{d}{2}} |\widehat{\operatorname{div} \mathbf{q}}|_{k,\widehat{K}} && \text{(by (1.6))} \\
&\quad \lesssim h_K^{k+1} \big(|u|_{k+1,K} + \tau_{\max}^{-1} |\operatorname{div} \mathbf{q}|_{k,K} \big). && \text{(by (1.9))}
\end{aligned}$$

and consequently

$$\begin{aligned}
&\|\widehat{u} - \widehat{\Pi}^{\mathrm{HDG}} \widehat{u}\|_{\partial \widehat{K}} \\
&\quad \lesssim \|\widehat{u} - \widehat{\Pi}^{\mathrm{HDG}} \widehat{u}\|_{1,\widehat{K}} && \text{(trace theorem)} \\
&\quad \leqslant \|\widehat{u} - \widehat{\Pi}_k \widehat{u}\|_{1,\widehat{K}} + \|\widehat{\Pi}_k \widehat{u} - \widehat{\Pi}^{\mathrm{HDG}} \widehat{u}\|_{1,\widehat{K}} \\
&\quad \lesssim |\widehat{u}|_{k+1,\widehat{K}} + \|\widehat{\Pi}_k \widehat{u} - \widehat{\Pi}^{\mathrm{HDG}} \widehat{u}\|_{\widehat{K}} && \text{(compactness and finite dim)} \\
&\quad \lesssim |\widehat{u}|_{k+1,\widehat{K}} + \|\widehat{u} - \widehat{\Pi}^{\mathrm{HDG}} \widehat{u}\|_{\widehat{K}} \\
&\quad \lesssim h_K^{-\frac{d}{2}} h_K^{k+1} \big(|u|_{k+1,K} + \tau_{\max}^{-1} |\operatorname{div} \mathbf{q}|_{k,K} \big).
\end{aligned}$$

We then apply Proposition 3.4 with $\varepsilon := \widehat{u} - \widehat{\Pi}^{\mathrm{HDG}} \widehat{u}$, so that

$$\begin{aligned}
&\|\mathbf{q} - \Pi^{\mathrm{HDG}} \mathbf{q}\|_K \\
&\quad \approx h_K^{1-\frac{d}{2}} \|\widehat{\mathbf{q}} - \widehat{\Pi}^{\mathrm{HDG}} \widehat{\mathbf{q}}\|_{\widehat{K}} && \text{(by (1.7) \& (3.7))} \\
&\quad \lesssim h_K^{1-\frac{d}{2}} \big(|\widehat{\mathbf{q}}|_{k+1,\widehat{K}} + \|\breve{\tau}\|_{L^\infty(\partial \widehat{K} \setminus \widehat{e})} \|\widehat{u} - \Pi^{\mathrm{HDG}} \widehat{u}\|_{\partial \widehat{K}} \big) && \text{(by Prop. 3.4)} \\
&\quad \lesssim h_K^{k+1} |\mathbf{q}|_{k+1,K} + \tau^\star h_K^{\frac{d}{2}} \|\widehat{u} - \Pi^{\mathrm{HDG}} \widehat{u}\|_{\partial \widehat{K}} && \text{(by (1.9) \& (3.16))} \\
&\quad \lesssim h_K^{k+1} \big(|\mathbf{q}|_{k+1,K} + \tau^\star |u|_{k+1,K} + \tau^\star \tau_{\max}^{-1} |\operatorname{div} \mathbf{q}|_{k,K} \big) \\
&\quad \lesssim h_K^{k+1} \big(|\mathbf{q}|_{k+1,K} + \tau^\star |u|_{k+1,K} \big).
\end{aligned}$$

This completes the proof.

An important observation. The entire analysis holds if we change τ by $-\tau$ in the definition of the projection. This is equivalent to changing the orientation of the normal vector and, as such, to a simple change of signs in some terms in the right-hand sides of the decoupled problems (3.8).

3.4 Error Analysis

3.4.1 Energy Arguments

Error equations. We start by redefining the local projections: we take $(\boldsymbol{\Pi}\mathbf{q}, \Pi u)$ to be the local HDG projection and $\mathrm{P}u$ to be (again) the orthogonal projection onto M_h. The discrete errors are the same quantities that we defined in (2.30)

$$\boldsymbol{\varepsilon}_h^q := \boldsymbol{\Pi}\mathbf{q} - \mathbf{q}_h \in \mathbf{V}_h, \qquad \varepsilon_h^u := \Pi u - u_h \in W_h, \qquad \widehat{\varepsilon}_h^u := \mathrm{P}u - \widehat{u}_h \in M_h.$$

We will also consider the error in the fluxes:

$$\begin{aligned}
\widehat{\varepsilon}_h^q &:= \boldsymbol{\Pi}\mathbf{q}\cdot\mathbf{n} + \tau(\Pi u - \mathrm{P}u) - \big(\mathbf{q}_h\cdot\mathbf{n} + \tau(u_h - \widehat{u}_h)\big)\\
&= \boldsymbol{\varepsilon}_h^q\cdot\mathbf{n} + \tau(\varepsilon_h^u - \widehat{\varepsilon}_h^u)\\
&= \mathrm{P}(\mathbf{q}\cdot\mathbf{n}) - \big(\mathbf{q}_h\cdot\mathbf{n} + \tau(u_h - \widehat{u}_h)\big) && \text{(see (3.4c))}\\
&= \mathrm{P}(\mathbf{q}\cdot\mathbf{n}) - \widehat{\mathbf{q}}_h\cdot\mathbf{n}. && \text{(see (3.2))}
\end{aligned}$$

This is how HDG projections and HDG equations interact:

$$\begin{aligned}
(\kappa^{-1}\boldsymbol{\Pi}\mathbf{q}, \mathbf{r})_{\mathscr{T}_h} - (\Pi u, \operatorname{div}\mathbf{r})_{\mathscr{T}_h} + \langle \mathrm{P}u, \mathbf{r}\cdot\mathbf{n}\rangle_{\partial\mathscr{T}_h} &= (\kappa^{-1}(\boldsymbol{\Pi}\mathbf{q} - \mathbf{q}), \mathbf{r})_{\mathscr{T}_h}, && (3.17a)\\
(\operatorname{div}\boldsymbol{\Pi}\mathbf{q}, w)_{\mathscr{T}_h} + \langle \tau(\Pi u - \mathrm{P}u), w\rangle_{\partial\mathscr{T}_h} &= (f, w)_{\mathscr{T}_h}, && (3.17b)\\
\langle \boldsymbol{\Pi}\mathbf{q}\cdot\mathbf{n} + \tau(\Pi u - \mathrm{P}u), \mu_1\rangle_{\partial\mathscr{T}_h\setminus\Gamma} &= 0, && (3.17c)\\
\langle \mathrm{P}u, \mu_2\rangle_\Gamma &= \langle g, \mu_2\rangle_\Gamma, && (3.17d)
\end{aligned}$$

for all $(\mathbf{r}, w, \mu_1, \mu_2) \in \mathbf{V}_h \times W_h \times M_h^\circ \times M_h^\Gamma$. Note that we have used the weak commutativity property (3.6) in (3.17b). The **error equations** are the difference between the latter and the HDG Eqs. (3.1):

$$\begin{aligned}
(\kappa^{-1}\boldsymbol{\varepsilon}_h^q, \mathbf{r})_{\mathscr{T}_h} - (\varepsilon_h^u, \operatorname{div}\mathbf{r})_{\mathscr{T}_h} + \langle \widehat{\varepsilon}_h^u, \mathbf{r}\cdot\mathbf{n}\rangle_{\partial\mathscr{T}_h} &= (\kappa^{-1}(\boldsymbol{\Pi}\mathbf{q} - \mathbf{q}), \mathbf{r})_{\mathscr{T}_h}, && (3.18a)\\
(\operatorname{div}\boldsymbol{\varepsilon}_h^q, w)_{\mathscr{T}_h} + \langle \tau(\varepsilon_h^u - \widehat{\varepsilon}_h^u), w\rangle_{\partial\mathscr{T}_h} &= 0, && (3.18b)\\
\langle \boldsymbol{\varepsilon}_h^q\cdot\mathbf{n} + \tau(\varepsilon_h^u - \widehat{\varepsilon}_h^u), \mu_1\rangle_{\partial\mathscr{T}_h\setminus\Gamma} &= 0, && (3.18c)\\
\langle \widehat{\varepsilon}_h^u, \mu_2\rangle_\Gamma &= 0. && (3.18d)
\end{aligned}$$

Once again, these equations faithfully replicate the error equations for mixed methods: taking $\tau = 0$, we obtain the error equations for RT (2.29) and BDM (Section 2.4.4), albeit with new polynomial spaces and projections.

Energy estimate. Testing Eqs. (3.18) with $(\boldsymbol{\varepsilon}_h^u, \varepsilon_h^u, -\widehat{\varepsilon}_h^u, -\widehat{\varepsilon}_h^q)$ and adding the results, we obtain an **energy identity**

$$(\kappa^{-1}\boldsymbol{\varepsilon}_h^q, \boldsymbol{\varepsilon}_h^q)_{\mathscr{T}_h} + \langle \tau(\varepsilon_h^u - \widehat{\varepsilon}_h^u), \varepsilon_h^u - \widehat{\varepsilon}_h^u\rangle_{\partial\mathscr{T}_h} = (\kappa^{-1}(\boldsymbol{\Pi}\mathbf{q} - \mathbf{q}), \boldsymbol{\varepsilon}_h^q)_{\mathscr{T}_h}, \qquad (3.19)$$

and a corresponding **energy estimate** that uses a parameter-dependent seminorm

$$|\mu|_\tau := \langle \tau\,\mu, \mu\rangle_{\partial\mathscr{T}_h}^{1/2} = \Big(\sum_{K\in\mathscr{T}_h} \langle \tau\,\mu, \mu\rangle_{\partial K}\Big)^{1/2},$$

so that we have

$$\|\boldsymbol{\varepsilon}_h^q\|_{\kappa^{-1}}^2 + |\varepsilon_h^u - \widehat{\varepsilon}_h^u|_\tau^2 \leqslant \|\boldsymbol{\Pi}\mathbf{q} - \mathbf{q}\|_{\kappa^{-1}}^2. \tag{3.20}$$

An estimate for the flux. As we saw in Sect. 2.2.2 (right before proving (2.34)), we can bound

$$h_K^{\frac{1}{2}}\|\mathbf{p}\cdot\mathbf{n}\|_{\partial K} \lesssim \|\mathbf{p}\|_K \qquad \forall \mathbf{p} \in \mathscr{P}_k(K),$$

and therefore

$$\|\widehat{\varepsilon}_h^q\|_h \leqslant \|\boldsymbol{\varepsilon}_h^q\cdot\mathbf{n}\|_h + \|\tau(\varepsilon_h^u - \widehat{\varepsilon}_h^u)\|_h \lesssim \|\boldsymbol{\varepsilon}_h^q\|_\Omega + \max_{K\in\mathscr{T}_h} h_K^{1/2}\|\tau\|_{L^\infty(\partial K)}^{1/2}\,|\varepsilon_h^u - \widehat{\varepsilon}_h^u|_\tau,$$

which together with the energy estimate (3.20) yield our second estimate for HDG

$$\|\widehat{\varepsilon}_h^q\|_h \lesssim \big(1 + \max_{K\in\mathscr{T}_h} h_K^{1/2}\|\tau\|_{L^\infty(\partial K)}^{1/2}\big)\|\boldsymbol{\Pi}\mathbf{q} - \mathbf{q}\|_{\kappa^{-1}}. \tag{3.21}$$

Bound for $\widehat{\varepsilon}_h^u$. If $k \geqslant 1$, we can locally lift the value $\widehat{\varepsilon}_h^u$ using the BDM lifting of Proposition 2.10. The arguments used to prove (2.84) are still valid, since they rely on the existence of a local lifting to the test space $\mathbf{V}_h$ and on the first error equation. We thus have

$$\|\widehat{\varepsilon}_h^u\|_h \lesssim \|\varepsilon_h^u\|_\Omega + h\|\mathbf{q} - \mathbf{q}_h\|_{\kappa^{-1}} \qquad \text{for } k \geqslant 1. \tag{3.22}$$

This argument will guarantee superconvergence of $\widehat{u}_h$ to $\mathrm{P}u$ whenever u_h superconverges to Πu. This will be the goal of the next section.

3.4.2 Duality Arguments

Estimates by duality arguments. In order to avoid some lengthy computations that have appeared in previous treatments of the duality arguments, we will use the more systematic approach of Sect. 2.3.5. The first step is the consideration of a dual problem:

$$\kappa^{-1}\boldsymbol{\xi} - \nabla\theta = \mathbf{0} \qquad \text{in } \Omega, \tag{3.23a}$$

$$-\mathrm{div}\,\boldsymbol{\xi} = \varepsilon_h^u \qquad \text{in } \Omega, \tag{3.23b}$$

$$\theta = 0 \qquad \text{on } \Gamma. \tag{3.23c}$$

Since the balance of signs between $\boldsymbol{\xi}$ and θ has changed, we will define $(\boldsymbol{\Pi}\boldsymbol{\xi}, \Pi\theta)$ to be the HDG projection corresponding to $-\tau$ (see the last comment of Sect. 3.3). We now write some equations satisfied by the projections, namely, what we would get in (3.17) if we modified the original equations to (3.23)

$$
\begin{aligned}
&(\kappa^{-1}\boldsymbol{\Pi}\boldsymbol{\xi}, \mathbf{r})_{\mathscr{T}_h} + (\Pi\theta, \operatorname{div}\mathbf{r})_{\mathscr{T}_h} - \langle \mathrm{P}\theta, \mathbf{r}\cdot\mathbf{n}\rangle_{\partial\mathscr{T}_h} &&= (\kappa^{-1}(\boldsymbol{\Pi}\boldsymbol{\xi} - \boldsymbol{\xi}), \mathbf{r})_{\mathscr{T}_h} && \forall \mathbf{r} \in \mathbf{V}_h,\\
&-(\operatorname{div}\boldsymbol{\Pi}\boldsymbol{\xi}, w)_{\mathscr{T}_h} + \langle \tau(\Pi\theta - \mathrm{P}\theta), w\rangle_{\partial\mathscr{T}_h} &&= (\varepsilon_h^u, w)_{\mathscr{T}_h} && \forall w \in W_h,\\
&\langle \boldsymbol{\Pi}\boldsymbol{\xi}\cdot\mathbf{n} - \tau(\Pi\theta - \mathrm{P}\theta), \mu\rangle_{\partial\mathscr{T}_h\setminus\Gamma} &&= 0 && \forall \mu \in M_h^\circ,\\
&\langle \mathrm{P}\theta, \mu\rangle_\Gamma &&= 0 && \forall \mu \in M_h^\Gamma.
\end{aligned}
$$

Note how the second equation—the weak commutativity property—and the third equation—the action of the projection on faces—have changed signs because of the fact that we are using $-\tau$ instead of τ. We next go ahead and test with the errors of the solution to the HDG equation. We are going to align everything in a careful way, since we want to add by columns instead of by rows:

$$
\begin{aligned}
&(\kappa^{-1}\boldsymbol{\Pi}\boldsymbol{\xi}, \boldsymbol{\varepsilon}_h^q)_{\mathscr{T}_h} + (\Pi\theta, \operatorname{div}\boldsymbol{\varepsilon}_h^q)_{\mathscr{T}_h} - \langle \mathrm{P}\theta, \boldsymbol{\varepsilon}_h^q\cdot\mathbf{n}\rangle_{\partial\mathscr{T}_h\setminus\Gamma} = (\kappa^{-1}(\boldsymbol{\Pi}\boldsymbol{\xi} - \boldsymbol{\xi}), \boldsymbol{\varepsilon}_h^q)_{\mathscr{T}_h},\\
&-(\operatorname{div}\boldsymbol{\Pi}\boldsymbol{\xi}, \varepsilon_h^u)_{\mathscr{T}_h} + \langle \tau\Pi\theta, \varepsilon_h^u\rangle_{\partial\mathscr{T}_h} - \langle \tau\mathrm{P}\theta, \varepsilon_h^u\rangle_{\partial\mathscr{T}_h\setminus\Gamma} = \|\varepsilon_h^u\|_\Omega^2,\\
&\langle \boldsymbol{\Pi}\boldsymbol{\xi}\cdot\mathbf{n}, \widehat{\varepsilon}_h^u\rangle_{\partial\mathscr{T}_h} - \langle \tau\Pi\theta, \widehat{\varepsilon}_h^u\rangle_{\partial\mathscr{T}_h} + \langle \tau\mathrm{P}\theta, \widehat{\varepsilon}_h^u\rangle_{\partial\mathscr{T}_h\setminus\Gamma} = 0.
\end{aligned}
$$

In between, we have used that $\widehat{\varepsilon}_h^u = 0$ on Γ (this was the fourth error Eq. (3.18d)) and $\mathrm{P}\theta = 0$ on Γ. We now sum these three equalities, but organize terms by column:

$$
\begin{aligned}
\|\varepsilon_h^u\|_\Omega^2 &+ (\kappa^{-1}(\boldsymbol{\Pi}\boldsymbol{\xi} - \boldsymbol{\xi}), \boldsymbol{\varepsilon}_h^q)_{\mathscr{T}_h}\\
&= (\kappa^{-1}\boldsymbol{\Pi}\boldsymbol{\xi}, \boldsymbol{\varepsilon}_h^q)_{\mathscr{T}_h} - (\operatorname{div}\boldsymbol{\Pi}\boldsymbol{\xi}, \varepsilon_h^u)_{\mathscr{T}_h} + \langle \boldsymbol{\Pi}\boldsymbol{\xi}\cdot\mathbf{n}, \widehat{\varepsilon}_h^u\rangle_{\partial\mathscr{T}_h}\\
&\quad + (\Pi\theta, \operatorname{div}\boldsymbol{\varepsilon}_h^q)_{\mathscr{T}_h} + \langle \tau\Pi\theta, \varepsilon_h^u\rangle_{\partial\mathscr{T}_h} - \langle \tau\Pi\theta, \widehat{\varepsilon}_h^u\rangle_{\partial\mathscr{T}_h}\\
&\quad - \langle \mathrm{P}\theta, \boldsymbol{\varepsilon}_h^q\cdot\mathbf{n}\rangle_{\partial\mathscr{T}_h\setminus\Gamma} - \langle \tau\mathrm{P}\theta, \varepsilon_h^u\rangle_{\partial\mathscr{T}_h\setminus\Gamma} + \langle \tau\mathrm{P}\theta, \widehat{\varepsilon}_h^u\rangle_{\partial\mathscr{T}_h\setminus\Gamma}\\
&= (\boldsymbol{\Pi}\boldsymbol{\xi}, \kappa^{-1}\boldsymbol{\varepsilon}_h^q)_{\mathscr{T}_h} - (\operatorname{div}\boldsymbol{\Pi}\boldsymbol{\xi}, \varepsilon_h^u)_{\mathscr{T}_h} + \langle \boldsymbol{\Pi}\boldsymbol{\xi}\cdot\mathbf{n}, \widehat{\varepsilon}_h^u\rangle_{\partial\mathscr{T}_h}\\
&\quad + (\Pi\theta, \operatorname{div}\boldsymbol{\varepsilon}_h^q)_{\mathscr{T}_h} + \langle \Pi\theta, \tau(\varepsilon_h^u - \widehat{\varepsilon}_h^u)\rangle_{\partial\mathscr{T}_h}\\
&\quad - \langle \mathrm{P}\theta, \boldsymbol{\varepsilon}_h^q\cdot\mathbf{n} + \tau(\varepsilon_h^u - \widehat{\varepsilon}_h^u)\rangle_{\partial\mathscr{T}_h\setminus\Gamma}\\
&= (\boldsymbol{\Pi}\boldsymbol{\xi}, \kappa^{-1}(\boldsymbol{\Pi}\mathbf{q} - \mathbf{q}))_{\mathscr{T}_h},
\end{aligned}
$$

where we have used the error Eqs. (3.18). What is left is a simple reorganization of terms in the above equality:

$$
\begin{aligned}
\|\varepsilon_h^u\|_\Omega^2 &= (\boldsymbol{\Pi}\boldsymbol{\xi}, \kappa^{-1}(\boldsymbol{\Pi}\mathbf{q} - \mathbf{q}))_{\mathscr{T}_h} - (\boldsymbol{\Pi}\boldsymbol{\xi} - \boldsymbol{\xi}, \kappa^{-1}(\boldsymbol{\Pi}\mathbf{q} - \mathbf{q}_h))_{\mathscr{T}_h}\\
&= (\boldsymbol{\Pi}\boldsymbol{\xi} - \boldsymbol{\xi}, \kappa^{-1}(\mathbf{q}_h - \mathbf{q}))_{\mathscr{T}_h} + (\boldsymbol{\xi}, \kappa^{-1}(\boldsymbol{\Pi}\mathbf{q} - \mathbf{q}))_{\mathscr{T}_h} && (\pm\boldsymbol{\xi})\\
&= (\boldsymbol{\Pi}\boldsymbol{\xi} - \boldsymbol{\xi}, \kappa^{-1}(\mathbf{q}_h - \mathbf{q}))_{\mathscr{T}_h} + (\nabla\theta, \boldsymbol{\Pi}\mathbf{q} - \mathbf{q})_{\mathscr{T}_h} && \text{(by (3.23a))}\\
&= (\boldsymbol{\Pi}\boldsymbol{\xi} - \boldsymbol{\xi}, \kappa^{-1}(\mathbf{q}_h - \mathbf{q}))_{\mathscr{T}_h} + (\nabla\theta - \boldsymbol{\Pi}_{k-1}\nabla\theta, \boldsymbol{\Pi}\mathbf{q} - \mathbf{q})_{\mathscr{T}_h}.
\end{aligned}
$$

Let us write this as an inequality:

$$\begin{aligned}\|\varepsilon_h^u\|_\Omega^2 \leqslant& \|\boldsymbol{\Pi}\boldsymbol{\xi} - \boldsymbol{\xi}\|_{\mathscr{T}_h} \|\kappa^{-1/2}\|_{L^\infty} \|\mathbf{q}_h - \mathbf{q}\|_{\kappa^{-1}} \\ &+ \|\nabla\theta - \boldsymbol{\Pi}_{k-1}\nabla\theta\|_{\mathscr{T}_h} \|\kappa^{1/2}\|_{L^\infty} \|\boldsymbol{\Pi}\mathbf{q} - \mathbf{q}\|_{\kappa^{-1}} \\ \lesssim& (\|\boldsymbol{\Pi}\boldsymbol{\xi} - \boldsymbol{\xi}\|_{\mathscr{T}_h} + \|\nabla\theta - \boldsymbol{\Pi}_{k-1}\nabla\theta\|_{\mathscr{T}_h}) \|\boldsymbol{\Pi}\mathbf{q} - \mathbf{q}\|_{\kappa^{-1}}.\end{aligned}$$

Assuming regularity

$$\|\boldsymbol{\xi}\|_{1,\Omega} + \|\theta\|_{2,\Omega} \leqslant C_{\text{reg}} \|\varepsilon_h^u\|_\Omega$$

for the solution of (3.23), the above argument leads to

$$\|\varepsilon_h^u\|_\Omega \lesssim h^{\min\{k,1\}} \|\boldsymbol{\Pi}\mathbf{q} - \mathbf{q}\|_{\kappa^{-1}}, \tag{3.24}$$

and hence to superconvergence when $k \geqslant 1$. For $k = 0$, no regularity of the dual problem is needed.

Wrap-up paragraph. The previous estimates together with the already studied properties of the HDG projection and of the projection P give the following table of convergence orders for smooth solutions.

$$\begin{aligned}
\|\mathbf{q} - \mathbf{q}_h\|_\Omega &\lesssim h^{k+1}, & \|\boldsymbol{\Pi}\mathbf{q} - \mathbf{q}_h\|_\Omega &\lesssim h^{k+1}, && \text{(see (3.20))}\\
\|u - u_h\|_\Omega &\lesssim h^{k+1}, & \|\Pi u - u_h\|_\Omega &\lesssim h^{k+1+\min\{k,1\}}, && \text{(see (3.24))}\\
\|u - \widehat{u}_h\|_h &\lesssim h^{k+1}, & \|\mathrm{P}u - \widehat{u}_h\|_h &\lesssim h^{k+2}, && (k \geqslant 1, \text{ see (3.22)})\\
\|\mathbf{q}\cdot\mathbf{n} - \widehat{\mathbf{q}}_h\cdot\mathbf{n}\|_h &\lesssim h^{k+1}, & \|\mathrm{P}(\mathbf{q}\cdot\mathbf{n}) - \widehat{\mathbf{q}}_h\cdot\mathbf{n}\|_h &\lesssim h^{k+1}. && \text{(see (3.21))}
\end{aligned}$$

3.5 HDG for the Helmholtz Equation

In this section, we will be dealing with HDG discretization of the Helmholtz equation. This requires the repeated use of complex-valued functions. To avoid confusion *all the brackets will be linear in both components*, and therefore, conjugates will have to be produced explicitly when needed. Sobolev spaces and polynomial spaces will be considered to be taken for functions with complex values. We will present an analysis for a problem with a fixed frequency (wave number) and for which there are no eigenvalues.

First order in space, second order in frequency formulation. We start with the Helmholtz equation written in second-order form. The coefficients are strongly positive functions $\rho, \kappa \in L^\infty(\Omega)$, $\rho \geqslant \rho_0 > 0$, $\kappa \geqslant \kappa_0 > 0$ and a positive parameter $\omega > 0$, playing the role of wave number. To add some more variety to the pool, we will use first-order absorbing boundary conditions (that is, wave number-dependent Robin boundary conditions):

$$-\mathrm{div}\,(\kappa\nabla u) - \omega^2 \rho\, u = f \qquad \text{in } \Omega, \tag{3.25a}$$
$$\kappa\nabla u\cdot\mathbf{n} - \imath\omega u = g \qquad \text{on } \Gamma. \tag{3.25b}$$

As usual, we introduce

$$\mathbf{q} := -\kappa\nabla u$$

to reach a first order in space and second order in frequency formulation

$$\kappa^{-1}\mathbf{q} + \nabla u = \mathbf{0} \qquad \text{in } \Omega, \tag{3.26a}$$
$$\mathrm{div}\,\mathbf{q} - \omega^2\rho\, u = f \qquad \text{in } \Omega, \tag{3.26b}$$
$$-\mathbf{q}\cdot\mathbf{n} - \imath\,\omega u = g \qquad \text{on } \Gamma, \tag{3.26c}$$

which is the one we will discretize.

HDG approximation. For $k \geqslant 0$, we consider the spaces

$$\mathbf{V}_h := \prod_{K\in\mathscr{T}_h} \mathscr{P}_k(K;\mathbb{C}), \qquad W_h := \prod_{K\in\mathscr{T}_h} \mathscr{P}_k(K;\mathbb{C}), \qquad M_h := \prod_{e\in\mathscr{E}_h} \mathscr{P}_k(e;\mathbb{C}).$$

The decomposition into interior and boundary parts for the space on the skeleton, $M_h = M_h^\circ \oplus M_h^\Gamma$, will not be used in the formulation of the method, since we are not dealing with Dirichlet conditions. As usual, we consider the stabilization function

$$\tau \in \prod_{K\in\mathscr{T}_h} \mathscr{R}_0(\partial K), \qquad \tau \geqslant 0 \qquad \tau|_{\partial K} \neq 0 \ \ \forall K.$$

The numerical flux (compare with (3.2)) is wave number (frequency)-dependent, and complex-valued

$$\widehat{\mathbf{q}}_h\cdot\mathbf{n} = \mathbf{q}_h\cdot\mathbf{n} - \imath\,\omega\,\tau\,(u_h - \widehat{u}_h) \in \mathscr{R}_k(\partial K) \qquad \forall K.$$

This mimics the boundary condition (3.26c) with the opposite sign. We will discuss this choice later. We look for

$$(\mathbf{q}_h, u_h, \widehat{u}_h) \in \mathbf{V}_h \times W_h \times M_h, \tag{3.27a}$$

satisfying

$$(\kappa^{-1}\mathbf{q}_h, \mathbf{r})_{\mathscr{T}_h} - (u_h, \mathrm{div}\,\mathbf{r})_{\mathscr{T}_h} + \langle \widehat{u}_h, \mathbf{r}\cdot\mathbf{n}\rangle_{\partial\mathscr{T}_h} = 0, \tag{3.27b}$$
$$(\mathrm{div}\,\mathbf{q}_h, w)_{\mathscr{T}_h} - \omega^2(\rho\, u_h, w)_{\mathscr{T}_h} - \imath\omega\langle \tau(u_h - \widehat{u}_h), w\rangle_{\partial\mathscr{T}_h} = (f, w)_{\mathscr{T}_h}, \tag{3.27c}$$
$$-\langle \mathbf{q}_h\cdot\mathbf{n} - \imath\omega\,\tau(u_h - \widehat{u}_h), \mu\rangle_{\partial\mathscr{T}_h} - \imath\omega\langle \widehat{u}_h, \mu\rangle_\Gamma = \langle g, \mu\rangle_\Gamma, \tag{3.27d}$$

for all $(\mathbf{r}, w, \mu) \in \mathbf{V}_h \times W_h \times M_h$. As usual, we can decompose (3.27d) into two sets of equations

$$-\langle \widehat{\mathbf{q}}_h \cdot \mathbf{n}, \mu \rangle_{\partial \mathcal{T}_h \setminus \Gamma} = 0 \qquad \forall \mu \in M_h^\circ,$$
$$-\langle \widehat{\mathbf{q}}_h \cdot \mathbf{n} + \imath\, \omega \widehat{u}_h, \mu \rangle_\Gamma = \langle g, \mu \rangle_\Gamma \qquad \forall \mu \in M_h^\Gamma.$$

First order in space and frequency formulation. Instead of introducing $\mathbf{q}$, we can introduce a different unknown $\mathbf{q}^\star$ and rescale the data according to the frequency:

$$\mathbf{q}^\star := -\frac{1}{\imath\,\omega}\kappa \nabla u = \frac{1}{\imath\,\omega}\mathbf{q}, \qquad f^\star := \frac{1}{\imath\,\omega} f, \qquad g^\star := \frac{1}{\imath\omega} g.$$

This leads to the following formulation:

$$\begin{aligned} \imath\omega\kappa^{-1}\mathbf{q}^\star + \nabla u &= \mathbf{0} && \text{in } \Omega,\\ \operatorname{div}\mathbf{q}^\star + \imath\omega\,\rho\, u &= f^\star && \text{in } \Omega,\\ -\mathbf{q}^\star \cdot \mathbf{n} - u &= g^\star && \text{on } \Gamma. \end{aligned}$$

For discretization purposes, the numerical flux

$$\widehat{\mathbf{q}}_h^\star \cdot \mathbf{n} = \mathbf{q}_h^\star \cdot \mathbf{n} - \tau\,(u_h - \widehat{u}_h)$$

is made to be independent of the frequency. The corresponding discrete equations are

$$\begin{aligned} \imath\omega(\kappa^{-1}\mathbf{q}_h^\star, \mathbf{r})_{\mathcal{T}_h} - (u_h, \operatorname{div}\mathbf{r})_{\mathcal{T}_h} + \langle \widehat{u}_h, \mathbf{r}\cdot\mathbf{n}\rangle_{\partial\mathcal{T}_h} &= 0 && \forall \mathbf{r} \in \mathbf{V}_h,\\ (\operatorname{div}\mathbf{q}_h^\star, w)_{\mathcal{T}_h} + \imath\omega(\rho\, u_h, w)_{\mathcal{T}_h} - \langle \tau(u_h - \widehat{u}_h), w\rangle_{\partial\mathcal{T}_h} &= (f^\star, w)_{\mathcal{T}_h} && \forall w \in W_h,\\ -\langle \mathbf{q}_h^\star \cdot \mathbf{n} - \tau(u_h - \widehat{u}_h), \mu\rangle_{\partial\mathcal{T}_h} - \langle \widehat{u}_h, \mu\rangle_\Gamma &= \langle g^\star, \mu\rangle_\Gamma && \forall \mu \in M_h. \end{aligned}$$

It is clear that the solutions of the above and of (3.27) are related by writing $\mathbf{q}_h = \imath\omega\mathbf{q}_h^\star$. The existing literature also allows for definitions of the numerical flux $\widehat{\mathbf{q}}_h^\star \cdot \mathbf{n}$ that depends on the frequency. This leads to a different class of methods, with essentially the same convergence properties for a fixed wave number ω. For instance, we can define

$$\widehat{\mathbf{q}}_h^\star \cdot \mathbf{n} = \mathbf{q}_h^\star \cdot \mathbf{n} \pm \imath\omega\tau\,(u_h - \widehat{u}_h),$$

which can be rewritten in a form similar to (3.27) with real penalization

$$\widehat{\mathbf{q}}_h \cdot \mathbf{n} = \mathbf{q}_h \cdot \mathbf{n} \mp \omega^2\,\tau\,(u_h - \widehat{u}_h).$$

The analysis of the resulting methods is not essentially different from what we will show in the next section.

3.5.1 Projection-Assisted Analysis

Complex HDG projections. Since the associated projection is expected to mimic the expression of the numerical flux, we need to redefine it. To do that, we introduce a family of complex HDG projections. Given $(\mathbf{q}, u) : K \to \mathbb{C}^d \times \mathbb{C}$ and $\beta \in \{1, -1, \imath, -\imath\}$, we define

$$(\boldsymbol{\Pi}^\beta \mathbf{q}, \Pi^\beta u) := (\boldsymbol{\Pi}_q^\beta(\mathbf{q}, u), \Pi_u^\beta(\mathbf{q}, u)) \in \boldsymbol{\mathcal{P}}_k(K;\mathbb{C}) \times \mathcal{P}_k(K;\mathbb{C})$$

as the solution to the equations

$$(\boldsymbol{\Pi}^\beta \mathbf{q}, \mathbf{r})_K = (\mathbf{q}, \mathbf{r})_K \qquad \forall \mathbf{r} \in \boldsymbol{\mathcal{P}}_{k-1}(K;\mathbb{C}), \tag{3.28a}$$

$$(\Pi^\beta u, v)_K = (u, v)_K \qquad \forall v \in \mathcal{P}_{k-1}(K;\mathbb{C}), \tag{3.28b}$$

$$\langle \boldsymbol{\Pi}^\beta \mathbf{q} \cdot \mathbf{n} + \beta\,\tau \Pi^\beta u, \mu\rangle_{\partial K} = \langle \mathbf{q} \cdot \mathbf{n} + \beta\,\tau\, u, \mu\rangle_{\partial K} \qquad \forall \mu \in \mathcal{R}_k(\partial K;\mathbb{C}). \tag{3.28c}$$

Proposition 3.6 (Estimates for the complex HDG projection) *Given* $\mathbf{q}$, u, *and* $0 \neq \tau \in \mathcal{R}_0(\partial K)$, $\tau \geqslant 0$, *and* $\beta \in \{1, -1, \imath, -\imath\}$

$$\|u - \Pi^\beta u\|_K \lesssim h_K^{k+1}\big(|u|_{k+1,K} + \tau_{\max}^{-1}|\mathrm{div}\,\mathbf{q}|_{k,K}\big), \tag{3.29a}$$

$$\|\mathbf{q} - \boldsymbol{\Pi}^\beta \mathbf{q}\|_K \lesssim h_K^{k+1}\big(|\mathbf{q}|_{k+1,K} + \tau^\star |u|_{k+1,K}\big), \tag{3.29b}$$

with $\tau_{\max} := \|\tau\|_{L^\infty}$ *and* $\tau^\star := \|\tau\|_{L^\infty(\partial K \setminus e)}$, *where* $\tau|_e = \tau_{\max}$.

Proof The cases $\beta = \pm 1$ were handled in Proposition 3.5. Note that, even if the data are complex-valued these projections work separately on the real and imaginary parts.

The cases $\beta = \pm\imath$ can be done by separating real and imaginary parts. We will only detail the computations for $\beta = \imath$. The only equation where β appears is (3.28c), and it can be shown to be equivalent to the equations

$$\langle \mathrm{Re}\,\boldsymbol{\Pi}^\imath \mathbf{q} \cdot \mathbf{n} - \tau \mathrm{Im}\, \Pi^\imath u, \mu\rangle_{\partial K} = \langle \mathrm{Re}\,\mathbf{q} \cdot \mathbf{n} - \tau\, \mathrm{Im}\, u, \mu\rangle_{\partial K} \qquad \forall \mu \in \mathcal{R}_k(\partial K),$$
$$\langle \mathrm{Im}\,\boldsymbol{\Pi}^\imath \mathbf{q} \cdot \mathbf{n} + \tau \mathrm{Re}\, \Pi^\imath u, \mu\rangle_{\partial K} = \langle \mathrm{Im}\,\mathbf{q} \cdot \mathbf{n} + \tau \mathrm{Re}\, u, \mu\rangle_{\partial K} \qquad \forall \mu \in \mathcal{R}_k(\partial K).$$

It is thus easy to prove that

$$\begin{aligned}(\boldsymbol{\Pi}^\imath \mathbf{q}, \Pi^\imath u) =&(\boldsymbol{\Pi}_q^{-1}(\mathrm{Re}\,\mathbf{q}, \mathrm{Im}\,u), \boldsymbol{\Pi}_q^{-1}(\mathrm{Re}\,\mathbf{q}, \mathrm{Im}\,u))\\ &+ \imath\,(\boldsymbol{\Pi}_q^{+1}(\mathrm{Im}\,\mathbf{q}, \mathrm{Re}\,u), \boldsymbol{\Pi}_q^{+1}(\mathrm{Im}\,\mathbf{q}, \mathrm{Re}\,u)).\end{aligned}$$

The estimates are then a simple consequence of Proposition 3.5.

We will use the projection with $\beta = -\imath$ and $\omega\tau$ in place of τ, which only requires a straightforward rescaling in Proposition 3.6. To make it more explicit, the equation (3.28c) is now

$$\langle \boldsymbol{\Pi}\mathbf{q}\cdot\mathbf{n} - \imath\omega\tau\,\Pi u, \mu\rangle_{\partial K} = \langle \mathbf{q}\cdot\mathbf{n} - \imath\omega\tau\, u, \mu\rangle_{\partial K} \qquad \forall \mu \in \mathscr{R}_k(\partial K).$$

A discrete Gårding inequality. The error quantities

$$\boldsymbol{\varepsilon}_h^q := \boldsymbol{\Pi}\mathbf{q} - \mathbf{q}_h \in \mathbf{V}_h, \qquad \varepsilon_h^u := \Pi u - u_h \in W_h, \qquad \widehat{\varepsilon}_h^u := \mathrm{P}u - \widehat{u}_h \in M_h,$$

will now be complemented with two approximation errors to make the analysis look more compact

$$\mathbf{e}_h^q := \boldsymbol{\Pi}\mathbf{q} - \mathbf{q}, \qquad e_h^u := \Pi u - u.$$

The error equations are now

$$(\kappa^{-1}\boldsymbol{\varepsilon}_h^q, \mathbf{r})_{\mathscr{T}_h} - (\varepsilon_h^u, \operatorname{div}\mathbf{r})_{\mathscr{T}_h} + \langle\widehat{\varepsilon}_h^u, \mathbf{r}\cdot\mathbf{n}\rangle_{\partial\mathscr{T}_h} = (\kappa^{-1}\mathbf{e}_h^q, \mathbf{r})_{\mathscr{T}_h}, \tag{3.30a}$$

$$(\operatorname{div}\boldsymbol{\varepsilon}_h^q, w)_{\mathscr{T}_h} - \omega^2(\rho\,\varepsilon_h^u, w)_{\mathscr{T}_h} - \imath\omega\langle\tau(\varepsilon_h^u - \widehat{\varepsilon}_h^u), w\rangle_{\partial\mathscr{T}_h} = -\omega^2(\rho\, e_h^u, w)_{\mathscr{T}_h}, \tag{3.30b}$$

$$-\langle\boldsymbol{\varepsilon}_h^q\cdot\mathbf{n} - \imath\omega\tau(\varepsilon_h^u - \widehat{\varepsilon}_h^u), \mu\rangle_{\partial\mathscr{T}_h} - \imath\omega\langle\widehat{\varepsilon}_h^u, \mu\rangle_\Gamma = 0, \tag{3.30c}$$

for all $(\mathbf{r}, w, \mu) \in \mathbf{V}_h \times W_h \times M_h$. In comparison with the error equations for diffusive problems, there are a couple of novelties: (a) the second error equation contains an error term associated to ε_h^u (this one would also appear if we were dealing with reaction–diffusion equations, but here it is going to be trouble because of the wrong sign); (b) the error equation (3.30a) associated to the "material law," i.e., to the equation where we write $\mathbf{q}$ in terms of u, will need to be conjugated for the error estimates (this equation plays a different role in a way); (c) there is a new boundary term in (3.30c), due to the impedance boundary condition. We will be using two norms for the forthcoming estimates:

$$\|u\|_\rho^2 := (\rho\, u, \overline{u})_{\mathscr{T}_h}, \qquad \|\mu\|_\Gamma^2 := \langle\mu, \overline{\mu}\rangle_\Gamma^2.$$

Instead of an energy estimate as in diffusion problems, we now obtain a Gårding-style identity. (Note how the equality of Proposition 3.7 is homogeneous in $\omega\, u$.) The proof of this result is very simple and left to the reader.

Proposition 3.7 *For the solution of the HDG Eqs.* (3.27)*, we have*

$$\|\boldsymbol{\varepsilon}_h^q\|_{\kappa^{-1}}^2 - \omega^2\|\varepsilon_h^u\|_\rho^2 - \imath\omega|\varepsilon_h^u - \widehat{\varepsilon}_h^u|_\tau^2 - \imath\omega\|\widehat{\varepsilon}_h^u\|_\Gamma^2 = (\kappa^{-1}\mathbf{e}_h^q, \overline{\boldsymbol{\varepsilon}_h^q})_{\mathscr{T}_h} - \omega^2(\rho\, e_h^u, \overline{\varepsilon_h^u})_{\mathscr{T}_h}.$$

Adjoint error estimates. In principle, we cannot obtain a direct estimate of any quantity from Proposition 3.7, and we will need to first obtain some estimates for the adjoint problem and then apply a bootstrapping argument. The adjoint equation, fed with the error ε_h^u as usual, is

$$\kappa^{-1}\boldsymbol{\xi} - \nabla\theta = \mathbf{0} \qquad \text{in } \Omega, \tag{3.31a}$$
$$-\operatorname{div}\boldsymbol{\xi} - \omega^2\rho\,\theta = \overline{\varepsilon_h^u} \qquad \text{in } \Omega, \tag{3.31b}$$
$$-\boldsymbol{\xi}\cdot\mathbf{n} + \imath\omega\theta = 0 \qquad \text{on } \Gamma, \tag{3.31c}$$

or in equivalent second-order form

$$-\operatorname{div}(\kappa\nabla\theta) - \omega^2\,\rho\,\theta = \overline{\varepsilon_h^u} \qquad \text{in } \Omega, \tag{3.32a}$$
$$-\kappa\nabla\theta\cdot\mathbf{n} + \imath\omega\theta = 0 \qquad \text{on } \Gamma. \tag{3.32b}$$

We assume that $(\theta, \boldsymbol{\xi})$ is smooth enough so that we can apply the HDG projection. This time we use $\beta = \imath$ and $\omega\tau$ instead of τ. There is a small difference in how we will treat the adjoint error equations. Instead of subtracting a $\mathscr{P}_{k-1}$ projection, we will subtract the average, i.e., the $\mathscr{P}_0$ projection. Let $[\rho\,\theta] \in \prod_{K\in\mathscr{T}_h}\mathscr{P}_0(K;\mathbb{C})$ and $[\kappa^{-1}\boldsymbol{\xi}] \in \prod_{K\in\mathscr{T}_h}\boldsymbol{\mathscr{P}}_0(K;\mathbb{C})$ be given by

$$[\rho\,\theta]_K := \begin{cases} \frac{1}{|K|}\int_K \rho\,\theta, & \text{if } k \geqslant 1, \\ 0, & \text{if } k = 0, \end{cases} \qquad [\kappa^{-1}\boldsymbol{\xi}]_K := \begin{cases} \frac{1}{|K|}\int_K \kappa^{-1}\boldsymbol{\xi}, & \text{if } k \geqslant 1, \\ 0, & \text{if } k = 0. \end{cases}$$

Proposition 3.8 *For the solution of the HDG Eqs.* (3.27), *we have*

$$\begin{aligned}\|\varepsilon_h^u\|_\Omega^2 =& \omega^2\left((\rho(\Pi\theta - \theta), \varepsilon_h^u - e_h^u)_{\mathscr{T}_h} - (e_h^u, \rho\theta - [\rho\theta])_{\mathscr{T}_h}\right) \\ &- \left((\kappa^{-1}(\boldsymbol{\Pi}\boldsymbol{\xi} - \boldsymbol{\xi}), \boldsymbol{\varepsilon}_h^q - \mathbf{e}_h^q)_{\mathscr{T}_h} - (\mathbf{e}_h^q, \kappa^{-1}\boldsymbol{\xi} - [\kappa^{-1}\boldsymbol{\xi}])_{\mathscr{T}_h}\right).\end{aligned} \tag{3.33}$$

Proof The equations satisfied by the adjoint projection on the solution of the adjoint problem, tested with the errors we are still trying to bound, are

$$\begin{aligned}&(\kappa^{-1}\boldsymbol{\Pi}\boldsymbol{\xi}, \boldsymbol{\varepsilon}_h^q)_{\mathscr{T}_h} + (\Pi\theta, \operatorname{div}\boldsymbol{\varepsilon}_h^q)_{\mathscr{T}_h} - \langle \mathrm{P}\theta, \boldsymbol{\varepsilon}_h^q\cdot\mathbf{n}\rangle_{\partial\mathscr{T}_h} \\ &\qquad = (\kappa^{-1}(\boldsymbol{\Pi}\boldsymbol{\xi} - \boldsymbol{\xi}), \boldsymbol{\varepsilon}_h^q)_{\mathscr{T}_h},\end{aligned} \tag{3.34a}$$

$$\begin{aligned}&-(\operatorname{div}\boldsymbol{\Pi}\boldsymbol{\xi}, \varepsilon_h^u)_{\mathscr{T}_h} - \omega^2(\rho\Pi\theta, \varepsilon_h^u)_{\mathscr{T}_h} - \imath\omega\langle\tau(\Pi\theta - \mathrm{P}\theta), \varepsilon_h^u\rangle_{\partial\mathscr{T}_h} \\ &\qquad = \|\varepsilon_h^u\|_\rho^2 - \omega^2(\rho(\Pi\theta - \theta), \varepsilon_h^u)_{\mathscr{T}_h},\end{aligned} \tag{3.34b}$$

$$\begin{aligned}&\langle\boldsymbol{\Pi}\boldsymbol{\xi}\cdot\mathbf{n} + \imath\omega\tau(\Pi\theta - \mathrm{P}\theta), \widehat{\varepsilon}_h^u\rangle_{\partial\mathscr{T}_h} - \imath\omega\langle\mathrm{P}\theta, \widehat{\varepsilon}_h^u\rangle_\Gamma \\ &\qquad = 0.\end{aligned} \tag{3.34c}$$

We then test the error Eqs. (3.30) with $(\boldsymbol{\Pi}\boldsymbol{\xi}, \Pi\theta, \mathrm{P}\theta)$, add, and compare with the sum of the Eqs. (3.34) to prove that

$$\|\varepsilon_h^u\|_\Omega^2 - \omega^2(\rho(\Pi\theta - \theta), \varepsilon_h^u)_{\mathcal{T}_h} + (\kappa^{-1}(\boldsymbol{\Pi\xi} - \boldsymbol{\xi}), \boldsymbol{\varepsilon}_h^q)_{\mathcal{T}_h} = (\kappa^{-1}\mathbf{e}_h^q, \boldsymbol{\Pi\xi})_{\mathcal{T}_h} - \omega^2(\rho\, e_h^u, \Pi\theta)_{\mathcal{T}_h}.$$

A simple rearrangement of terms yields

$$\begin{aligned}\|\varepsilon_h^u\|_\Omega^2 =&\omega^2\left((\rho(\Pi\theta - \theta), \varepsilon_h^u)_{\mathcal{T}_h} - (\rho\, e_h^u, \Pi\theta)_{\mathcal{T}_h}\right)\\ &- \left((\kappa^{-1}(\boldsymbol{\Pi\xi} - \boldsymbol{\xi}), \boldsymbol{\varepsilon}_h^q)_{\mathcal{T}_h} - (\kappa^{-1}\mathbf{e}_h^q, \boldsymbol{\Pi\xi})_{\mathcal{T}_h}\right).\end{aligned} \tag{3.35}$$

A small computation now shows

$$\begin{aligned}&(\rho(\Pi\theta - \theta), \varepsilon_h^u)_{\mathcal{T}_h} - (\rho\, e_h^u, \Pi\theta)_{\mathcal{T}_h}\\ &= (\rho(\Pi\theta - \theta), \varepsilon_h^u - e_h^u)_{\mathcal{T}_h} - (\rho\, e_h^u, \theta)_{\mathcal{T}_h}\\ &= (\rho(\Pi\theta - \theta), \varepsilon_h^u - e_h^u)_{\mathcal{T}_h} - (e_h^u, \rho\,\theta - [\rho\,\theta])_{\mathcal{T}_h}.\end{aligned}$$

The second group of terms in (3.35) can be handled similarly.

3.5.2 The Bootstrapping Argument

We are now ready to start with the actual error analysis, which will be obtained by combining Propositions 3.7 and 3.8. Note that each of these identities relies on right-hand sides that are handled in the other identity. This is where a common bootstrapping technique will come in handy. We start by assuming an elliptic regularity hypothesis

$$\|\theta\|_{1,\Omega} \leqslant C_\omega \|\varepsilon_h^u\|_\Omega, \qquad \|\boldsymbol{\xi}\|_{1,\Omega} \leqslant D_\omega \|\varepsilon_h^u\|_\Omega.$$

We will also assume that ρ and κ^{-1} are in $W^{1,\infty}$, at least on each of the elements. For simplicity, from now on $\tau_{\max} = 1$. Low-order estimates for the projection (use Proposition 3.6, note the scaling with ω, and take $\tau_{\max} = 1$) yield

$$\begin{aligned}\|\theta - \Pi\theta\|_\Omega &\lesssim h(|\theta|_{1,\Omega} + \omega^{-1}\|\operatorname{div}\boldsymbol{\xi}\|_\Omega)\\ &\lesssim h(C_\omega + \omega\, C_\omega + \omega^{-1})\|\varepsilon_h^u\|_\Omega,\end{aligned} \tag{3.36a}$$

$$\|\boldsymbol{\xi} - \boldsymbol{\Pi\xi}\|_\Omega \lesssim h(|\boldsymbol{\xi}|_{1,\Omega} + \omega\tau^\star|\theta|_{1,\Omega}) \lesssim h(D_\omega + \omega C_\omega \tau^\star)\|\varepsilon_h^u\|_\Omega, \tag{3.36b}$$

$$\|\rho\theta - [\rho\theta]\|_\Omega \lesssim \begin{cases} h|\rho\theta|_{1,\mathcal{T}_h} \lesssim C_\omega h\|\varepsilon_h^u\|_\Omega, & k \geqslant 1,\\ \|\theta\|_\Omega \lesssim C_\omega\|\varepsilon_h^u\|_\Omega, & k = 0,\end{cases} \tag{3.36c}$$

$$\|\kappa^{-1}\boldsymbol{\xi} - [\kappa^{-1}\boldsymbol{\xi}]\|_\Omega \lesssim \begin{cases} h|\kappa^{-1}\boldsymbol{\xi}|_{1,\mathcal{T}_h} \lesssim hD_\omega\|\varepsilon_h^u\|_\Omega, & k \geqslant 1,\\ \|\boldsymbol{\xi}\|_\Omega \lesssim C_\omega\|\varepsilon_h^u\|_\Omega, & k = 0.\end{cases} \tag{3.36d}$$

A neat trick to avoid being overwhelmed by notation is to play the bootstrapping argument as an algebraic computation. To do that we introduce the quantities:

$$Q := \|\boldsymbol{\varepsilon}_h^q\|_{\kappa^{-1}}, \qquad U := \omega\|\varepsilon_h^u\|_\rho, \qquad T := \omega^{1/2}(|\varepsilon_h^u - \widehat{\varepsilon}_h^u|_\tau^2 + \|\widehat{\varepsilon}_h^u\|_\Gamma^2)^{1/2}, \tag{3.37a}$$

$$E := \|\mathbf{e}_h^q\|_{\kappa^{-1}}, \qquad F := \omega\|e_h^u\|_\rho. \tag{3.37b}$$

We have proved in Proposition 3.7 that

$$(Q^2 + T^2) \leqslant 2|Q^2 - \imath T^2| \lesssim U^2 + E\,Q + F\,U.$$

We can thus apply Young's inequality here and simplify to

$$Q + T \lesssim U + E + F.$$

Proposition 3.8 and the projection estimates (3.36) yield

$$\begin{aligned}\omega^{-2}U^2 \lesssim & h(C_\omega + \omega C_\omega + \omega^{-1})U(U+F) + F\begin{Bmatrix} h \\ 1 \end{Bmatrix} C_\omega U \\ & + h(\omega^{-1}D_\omega + \tau^\star C_\omega)U(Q+E) + E\begin{Bmatrix} h\omega^{-1}D_\omega \\ \omega^{-1}C_\omega \end{Bmatrix} U. \end{aligned} \tag{3.38}$$

Analysis for $k \geqslant 1$. Let

$$\alpha := h\omega(\omega C_\omega + \omega^2 C_\omega + 1), \qquad \gamma := h(\omega D_\omega + \tau^\star\omega^2 C_\omega),$$

For fixed ω, we have $\alpha, \gamma = \mathscr{O}(h)$ and (3.38) reads

$$U \lesssim \alpha U + \alpha F + \gamma Q + \gamma E.$$

Overestimating $\alpha \lesssim 1$, we end up with two inequalities

$$Q + T \lesssim U + E + F, \qquad \text{(energy identity)} \tag{3.39a}$$

$$U \lesssim \alpha F + \gamma Q + \gamma E. \qquad \text{(duality argument)} \tag{3.39b}$$

We now plug the second of these inequalities into the first to obtain

$$Q + T \lesssim \alpha F + \gamma Q + \gamma E + E + F.$$

Overestimating again $\alpha + \gamma \lesssim 1$, we can apply Young's inequality and prove

$$Q + T \lesssim F + E = \mathscr{O}(h^{k+1}).$$

Therefore

$$U \lesssim (\alpha+\gamma)F + \gamma E = \mathcal{O}(h^{k+2}).$$

This proves optimal convergence of the method. Note that the bounds $\alpha + \gamma \lesssim 1$, for fixed ω, amount to nothing since they are absorbed with a fixed constant hidden in the $\lesssim$ symbol. However, if we keep ω variable, then h has to be small enough to compensate for the growth of the terms depending on ω, some of which depend on constants that are not easy to estimate.

The case $k = 0$. In this case, we will not have superconvergence, but we can still prove estimates. We define

$$\begin{aligned}
\alpha &:= h\omega(\omega C_\omega + \omega^2 C_\omega + 1),\\
\beta &:= \max(\omega^2, \omega)\, C_\omega,\\
\gamma &:= h(\omega D_\omega + \tau^\star \omega^2 C_\omega),
\end{aligned}$$

and note that, for fixed frequency, $\alpha, \gamma = \mathcal{O}(h)$ while $\beta = \mathcal{O}(1)$. We now have

$$U \lesssim \alpha U + \alpha F + \beta(F+E) + \gamma(Q+E).$$

We overestimate again $\alpha \lesssim 1$ and simplify

$$Q + T \lesssim U + E + F, \qquad \text{(energy identity)} \tag{3.40a}$$

$$U \lesssim \alpha F + \beta(F+E) + \gamma(Q+E). \qquad \text{(duality argument)} \tag{3.40b}$$

Bootstrapping, and simplifying, we first have

$$Q + T \lesssim (\alpha + \beta + 1)F + (\beta + \gamma + 1)E + \gamma Q.$$

Using $\gamma \lesssim 1$, we can apply Young's inequality and end up with

$$Q + T \lesssim F + E = \mathcal{O}(h^{k+1}), \qquad U \lesssim (\alpha+\beta+\gamma)F + (\beta+\gamma)E = \mathcal{O}(h^{k+1}),$$

because $\beta = \mathcal{O}(1)$. This completes the error analysis.

A word on unique solvability. In a certain way, we have cheated in the previous analysis, since we never got to prove that the discrete Eqs. (3.27) are uniquely solvable. We will develop here a very simple and clever idea of Cockburn to prove uniqueness from the error estimates. First of all, let us recall that (3.27) is a square system of linear equations, so uniqueness of solution is equivalent to the existence of solution for any given right-hand side. The analysis we have performed works for any solution $(\mathbf{q}_h, u_h, \widehat{u}_h)$ of the discrete Eqs. (3.27), provided that such a solution exists. Now, for $f = 0$ and $g = 0$, we know that Eqs. (3.27) are solvable, but maybe not uniquely. Let then $(\mathbf{q}_h, u_h, \widehat{u}_h)$ be any solution to (3.27) with homogeneous data. Since this is a discrete approximation of the exact solution $(\mathbf{q} = \mathbf{0}, u = 0)$ we can apply the error estimates above. We now have

$$\boldsymbol{\varepsilon}_h^q := -\mathbf{q}_h, \qquad \varepsilon_h^u := -u_h, \qquad \widehat{\varepsilon}_h^u := -\widehat{u}_h \in M_h, \qquad \mathbf{e}_h^q := \mathbf{0}, \qquad e_h^u := 0.$$

In a very condensed form, we can summarize the bootstrapping argument as the inequality $Q + T + U \lesssim E + F$ for h small enough. If we go back to the letter assignments (3.37), we can see that we have proved that $\mathbf{q}_h = \mathbf{0}$, $u_h = 0$ and $\tau \widehat{u}_h = 0$. The proof that $\widehat{u}_h = 0$ follows from $\langle \widehat{u}_h, \mathbf{r} \cdot \mathbf{n} \rangle_{\partial \mathscr{T}_h} = 0$ for all $\mathbf{r} \in \mathbf{V}_h$ (this is Eq. (3.27b) once we know that $\mathbf{q}_h$ and u_h vanish), with a simple argument based on Lemmas 2.1 and 2.2.

3.5.3 Local Solvability

In order for the HDG equations to actually be hybridizable, we need to show unique solvability of the local equations for each of the elements, i.e., we need to prove that if f and $\widehat{u}_h$ are given, we can solve for $\mathbf{q}_h$ and u_h elementwise, using (3.27b)–(3.27c). We start with two technical lemmas.

Lemma 3.1 *In the space* $\{u \in \mathscr{P}_k(K) : u|_e = 0 \text{ for some } e \in \mathscr{E}(K)\}$, *we have* $\|u\|_K \approx \|\Pi_{k-1} u\|_K$, *where* Π_{k-1} *is the orthogonal projection onto* $\mathscr{P}_{k-1}(K)$.

Proof The result is first proved in the reference element using Lemma 2.1(a) and then transferred to K with a simple scaling argument.

Lemma 3.2 *There exists a linear operator* $\mathrm{div}^+ : \mathscr{P}_{k-1}(K) \to \boldsymbol{\mathscr{P}}_k(K)$ *such that*

(a) $\mathrm{div}\,\mathrm{div}^+ u = u \quad \forall u \in \mathscr{P}_{k-1}(K)$,
(b) $\|\mathrm{div}^+ u\|_K \lesssim h_K \|u\|_K \quad \forall u \in \mathscr{P}_{k-1}(K)$.

Proof As usual, let us begin with work in the reference element. Take $\widehat{\mathrm{div}}^+$ to be the Moore–Penrose pseudoinverse of the surjective operator $\mathrm{div} : \boldsymbol{\mathscr{P}}_k(K) \to \mathscr{P}_{k-1}(K)$ and then define div^+ with the rule (recall the hat-check rules in Sect. 1.1)

$$\widehat{\mathrm{div}^+ u} := \widehat{\mathrm{div}}^+ \check{u}.$$

We thus have

$$\begin{aligned} \|\mathrm{div}^+ u\|_K &\approx h_K^{1-d/2} \|\widehat{\mathrm{div}^+ u}\|_{\widehat{K}} && \text{(by (1.7))} \\ &\lesssim h_K^{1-d/2} \|\check{u}\|_{\widehat{K}} && \text{(finite dimensionality)} \\ &\approx h_K \|u\|_K. && \text{(easy from bounds in Sect. 1.1)} \end{aligned}$$

This proves (b). To prove (a) note that

$$\widehat{\mathrm{div}\,\mathrm{div}^+ u} = \widehat{\mathrm{div}}\, \widehat{\mathrm{div}^+ u} = \widehat{\mathrm{div}}\, \widehat{\mathrm{div}}^+ \check{u} = \check{u},$$

as follows from (1.3a), the definition of div^+, and the fact that $\widehat{\mathrm{div}}^+$ is a right-inverse to the divergence operator on $\mathscr{P}_k(K)$.

Proposition 3.9 (Local solvability) *There exists a constant $C > 0$, depending on the shape-regularity constants of the mesh and on the polynomial degree k such that if*

$$h_K \omega C \|\rho\|^{1/2}_{L^\infty(K)} \|\kappa^{-1}\|^{1/2}_{L^\infty(K)} < 1, \tag{3.41}$$

then Eqs. (3.27b)–(3.27c), *counting* $(\mathbf{q}_h, u_h)$ *as unknowns, are uniquely solvable in* K. *Therefore, if* (3.41) *holds for every* $K \in \mathscr{T}_h$, *the system* (3.27) *is hybridizable.*

Proof To shorten some forthcoming expressions, we will write

$$\|\mathbf{q}_h\|_{\kappa^{-1},K} := \|\kappa^{-1/2}\mathbf{q}_h\|_K, \qquad \|u_h\|_{\rho,K} := \|\rho^{1/2}u_h\|_K.$$

We obviously only need to show that if (3.41) holds, the only solution to

$$(\kappa^{-1}\mathbf{q}_h, \mathbf{r})_K - (u_h, \mathrm{div}\,\mathbf{r})_K = 0 \qquad \forall \mathbf{r} \in \boldsymbol{\mathscr{P}}_k(K), \tag{3.42a}$$

$$(\mathrm{div}\,\mathbf{q}_h, w)_K - \omega^2(\rho\, u_h, w)_K - \imath\,\omega\langle \tau u_h, w\rangle_{\partial K} = 0 \qquad \forall w \in \mathscr{P}_k(K), \tag{3.42b}$$

is the homogeneous solution. Testing (3.42a) with $\mathbf{r} = \overline{\mathbf{q}_h}$ and (3.42b) with $w = \overline{u_h}$, it follows that

$$\|\mathbf{q}_h\|^2_{\kappa^{-1},K} - \omega^2\|u_h\|^2_{\rho,K} - \imath\omega\langle \tau u_h, \overline{u_h}\rangle_{\partial K} = 0$$

and therefore

$$\|\mathbf{q}_h\|^2_{\kappa^{-1},K} = \omega^2\|u_h\|^2_{\rho,K}, \qquad \tau u_h|_{\partial K} = 0. \tag{3.43}$$

Taking now $\mathbf{r} = \mathrm{div}^+\Pi_{k-1}\overline{u_h}$ in (3.42a) (see Lemmas 3.1 and 3.2), it follows that

$$\begin{aligned}
\|\Pi_{k-1}u_h\|^2_K =&(\kappa^{-1}\mathbf{q}_h, \mathrm{div}^+\Pi_{k-1}\overline{u_h})_K\\
\leqslant&\|\kappa^{-1}\|^{1/2}_{L^\infty(K)}\|\mathbf{q}_h\|_{\kappa^{-1},K}\|\mathrm{div}^+\Pi_{k-1}u_h\|_K\\
\leqslant&C_1 h_K\|\kappa^{-1}\|^{1/2}_{L^\infty(K)}\|\mathbf{q}_h\|_{\kappa^{-1},K}\|\Pi_{k-1}u_h\|_K. \qquad \text{(Lemma 3.2)}
\end{aligned}$$

Going back to (3.43) and using the constant $C_2 > 0$ hidden in Lemma 3.1, we have

$$\begin{aligned}
\|\mathbf{q}_h\|_{\kappa^{-1},K} =&\omega\|u_h\|_{\rho,K} \leqslant \omega\|\rho\|^{1/2}_{L^\infty(K)}\|u_h\|_K\\
\leqslant&C_2\omega\|\rho\|^{1/2}_{L^\infty(K)}\|\Pi_{k-1}u_h\|_K \qquad \text{(Lemma 3.1 since } \tau u_h = 0)\\
\leqslant&C_1C_2h_K\omega\|\rho\|^{1/2}_{L^\infty(K)}\|\kappa^{-1}\|^{1/2}_{L^\infty(K)}\|\mathbf{q}_h\|_{\kappa^{-1},K}.
\end{aligned}$$

If we take $C = C_1C_2$ and assume that (3.41) holds, then $\mathbf{q}_h = \mathbf{0}$ and therefore $\Pi_{k-1}u_h = 0$. However, $u_h = 0$ on a face of K at least (this follows from the fact that $\tau u_h = 0$ on ∂K) and therefore, by Lemma 2.1(a), $u_h = 0$. This completes the proof.

Exercises

1. Prove the error Eqs. (3.30).
2. Prove the energy estimate of Proposition 3.7. (Hint. Test with $\mathbf{r} = \overline{\boldsymbol{\varepsilon}_h^q}$, $w = \overline{\varepsilon_h^u}$, and $\mu = \overline{\widehat{\varepsilon}_h^u}$, conjugate the first equation, and add the results.)

Chapter 4
Variants of the HDG Method

4.1 The HDG+ Method and Its Projection

In this section, we introduce a variant of the HDG method with very much the same convergence properties as the original one, without the additional need of a postprocessing step. To distinguish the methods easily, we will refer to this scheme as the HDG+ method and refer to the previously studied scheme as the classical HDG method. This method involves a projection in the numerical flux, an idea that can be traced back to the work of Lehrenfeld and Schöberl [87]. In the following table, we compare the spaces for the hybridizable formulations of RT and BDM with HDG and the new variant of HDG that we will be discussing. Note that the stabilization parameter τ does not appear in the hybridizable formulation of the mixed methods (it can be taken to equal zero) and that the space for the variable on the skeleton $\widehat{u}_h$ is always the same: locally it is $\mathscr{R}_k(\partial K)$, and globally it is the space $M_h = \prod_{e\in\mathscr{E}_h} \mathscr{P}_k(e)$.

method	degree	$\mathbf{q}_h$	u_h	τ
RT	$k \geqslant 0$	$\mathscr{R}\mathscr{T}_k(K)$	$\mathscr{P}_k(K)$	0
BDM	$k \geqslant 1$	$\boldsymbol{\mathscr{P}}_k(K)$	$\mathscr{P}_{k-1}(K)$	0
HDG	$k \geqslant 0$	$\boldsymbol{\mathscr{P}}_k(K)$	$\mathscr{P}_k(K)$	$\mathscr{O}(1)$
HDG+	$k \geqslant 0$	$\boldsymbol{\mathscr{P}}_k(K)$	$\mathscr{P}_{k+1}(K)$	$\approx h_K^{-1}$

For $k \geqslant 0$, we consider the spaces

$$\mathbf{V}_h := \prod_{K\in\mathscr{T}_h} \boldsymbol{\mathscr{P}}_k(K), \quad W_h := \prod_{K\in\mathscr{T}_h} \mathscr{P}_{k+1}(K), \quad M_h := \prod_{e\in\mathscr{E}_h} \mathscr{P}_k(e), \tag{4.1}$$

S. Du and F.-J. Sayas, *An Invitation to the Theory of the Hybridizable Discontinuous Galerkin Method*, SpringerBriefs in Mathematics,
https://doi.org/10.1007/978-3-030-27230-2_4

and the subspace decomposition $M_h = M_h^\circ \oplus M_h^\Gamma$. Note how the polynomial degree for W_h is one order higher, which makes this method resemble a traditional Finite Element scheme, and quite the opposite of a BDM method (where $\mathbf{V}_h$ is the higher order space). The **stabilization function** now is of the form

$$\tau \in \prod_{K\in\mathscr{T}_h} \mathscr{R}_0(\partial K), \qquad \tau|_{\partial K} \approx h_K^{-1} \quad \forall K,$$

where the last condition can be written as the existence of two positive constants such that

$$c_1 h_K^{-1} \leqslant \tau|_{\partial K} \leqslant c_2 h_K^{-1} \quad \forall K.$$

A final ingredient of the method is the orthogonal projection P

$$\mathrm{P} : \prod_{K\in\mathscr{T}_h} L^2(\partial K) \to \prod_{K\in\mathscr{T}_h} \mathscr{R}_k(\partial K),$$

that we have used for theoretical purposes in past sections. This projection will be a key ingredient for the method. The new numerical flux is

$$\widehat{\mathbf{q}}_h \cdot \mathbf{n} := \mathbf{q}_h \cdot \mathbf{n} + \tau(\mathrm{P}u_h - \widehat{u}_h) \in \mathscr{R}_k(\partial K) \quad \forall K. \tag{4.2}$$

Once these ingredients have been introduced, we can easily define the new HDG scheme by slightly modifying the Eqs. (3.1). For easy reference, we will call this method the HDG+ scheme.

We look for

$$(\mathbf{q}_h, u_h, \widehat{u}_h) \in \mathbf{V}_h \times W_h \times M_h, \tag{4.3a}$$

satisfying

$$(\kappa^{-1}\mathbf{q}_h, \mathbf{r})_{\mathscr{T}_h} - (u_h, \operatorname{div}\mathbf{r})_{\mathscr{T}_h} + \langle \widehat{u}_h, \mathbf{r}\cdot\mathbf{n}\rangle_{\partial\mathscr{T}_h} = 0 \qquad \forall \mathbf{r} \in \mathbf{V}_h, \tag{4.3b}$$

$$(\operatorname{div}\mathbf{q}_h, w)_{\mathscr{T}_h} + \langle \tau(\mathrm{P}u_h - \widehat{u}_h), w\rangle_{\partial\mathscr{T}_h} = (f, w)_{\mathscr{T}_h} \qquad \forall w \in W_h, \tag{4.3c}$$

$$\langle \mathbf{q}_h\cdot\mathbf{n} + \tau(u_h - \widehat{u}_h), \mu\rangle_{\partial\mathscr{T}_h\setminus\Gamma} = 0 \qquad \forall \mu \in M_h^\circ, \tag{4.3d}$$

$$\langle \widehat{u}_h, \mu\rangle_\Gamma = \langle g, \mu\rangle_\Gamma \qquad \forall \mu \in M_h^\Gamma. \tag{4.3e}$$

Some comments. Once again, Eqs. (4.3b) and (4.3c) are local due to the fact that the spaces are discontinuous. After integration by parts, Eq. (4.3c) can be shown to be equivalent to

$$-(\mathbf{q}_h, \nabla w)_{\mathscr{T}_h} + \langle \widehat{\mathbf{q}}_h\cdot\mathbf{n}, w\rangle_{\partial\mathscr{T}_h} = (f, w)_{\mathscr{T}_h} \qquad \forall w \in W_h,$$

using the flux defined in (4.2). Equation (4.3d) can also be written as

$$\langle \mathbf{q}_h\cdot\mathbf{n} + \tau(\mathrm{P}u_h - \widehat{u}_h), \mu\rangle_{\partial\mathscr{T}_h\setminus\Gamma} = 0 \qquad \forall \mu \in M_h^\circ,$$

after noting that $\langle \tau \mathrm{P} u_h, \mu\rangle_{\partial\mathcal{T}_h} = \langle \tau u_h, \mu\rangle_{\partial\mathcal{T}_h}$ for all $\mu \in M_h$, since τ is piecewise constant and therefore $\tau\mu$ is piecewise $\mathcal{P}_k(e)$ on each $e \in \mathcal{E}_h$. Equivalently, we can write (4.3d) using the numerical flux (4.2)

$$\langle \widehat{\mathbf{q}}_h \cdot \mathbf{n}, \mu\rangle_{\partial\mathcal{T}_h\setminus\Gamma} = 0 \qquad \forall \mu \in M_h^\circ,$$

which shows that this condition is equivalent to $\widehat{\mathbf{q}}_h \cdot \mathbf{n}$ being single-valued. Note finally, that in the form (4.3), the projection P only makes an appearance in the symmetric positive semidefinite bilinear form

$$W_h \ni u_h, w \quad \longmapsto \quad \langle \tau \mathrm{P} u_h, w\rangle_{\partial\mathcal{T}_h} = \langle \tau u_h, \mathrm{P} w\rangle_{\partial\mathcal{T}_h} = \langle \tau \mathrm{P} u_h, \mathrm{P} w\rangle_{\partial\mathcal{T}_h}.$$

Proposition 4.1 (Unique solvability) *Equations* (4.3) *are uniquely solvable.*

Proof As usual, we only need to show that the only

$$(\mathbf{q}_h, u_h, \widehat{u}_h) \in \mathbf{V}_h \times W_h \times M_h,$$

satisfying

$$\begin{aligned}
&(\kappa^{-1}\mathbf{q}_h, \mathbf{r})_{\mathcal{T}_h} - (u_h, \operatorname{div}\mathbf{r})_{\mathcal{T}_h} + \langle \widehat{u}_h, \mathbf{r}\cdot\mathbf{n}\rangle_{\partial\mathcal{T}_h} &&= 0 \qquad \forall \mathbf{r} \in \mathbf{V}_h, && (4.4a)\\
&(\operatorname{div}\mathbf{q}_h, w)_{\mathcal{T}_h} + \langle \tau(\mathrm{P}u_h - \widehat{u}_h), \mathrm{P}w\rangle_{\partial\mathcal{T}_h} &&= 0 \qquad \forall w \in W_h, && (4.4b)\\
&-\langle \mathbf{q}_h\cdot\mathbf{n} + \tau(\mathrm{P}u_h - \widehat{u}_h), \mu\rangle_{\partial\mathcal{T}_h\setminus\Gamma} &&= 0 \qquad \forall \mu \in M_h^\circ, && (4.4c)\\
&\langle \widehat{u}_h, \mu\rangle_\Gamma &&= 0 \qquad \forall \mu \in M_h^\Gamma && (4.4d)
\end{aligned}$$

is the zero triplet. (Note that we have included additional occurrences of the projection P that leave the equations unchanged.) Equation (4.4d) is equivalent to $\widehat{u}_h = 0$ on Γ. Testing Eqs. (4.4a)–(4.4c) with $\mathbf{r} = \mathbf{q}_h$, $w = u_h$, $\mu = \widehat{u}_h$ and adding the equations, it follows that

$$(\kappa^{-1}\mathbf{q}_h, \mathbf{q}_h)_{\mathcal{T}_h} + \langle \tau(\mathrm{P}u_h - \widehat{u}_h), \mathrm{P}u_h - \widehat{u}_h\rangle_{\partial\mathcal{T}_h} = 0,$$

and therefore $\mathbf{q}_h = \mathbf{0}$ and $\widehat{u}_h = \mathrm{P}u_h$ on ∂K for all K. (Note that in this method $\tau > 0$.) Going back to (4.4a), we have

$$-(u_h, \operatorname{div}\mathbf{r})_{\mathcal{T}_h} + \langle u_h, \mathbf{r}\cdot\mathbf{n}\rangle_{\partial T_h} = 0 \qquad \forall \mathbf{r} \in \mathbf{V}_h,$$

(recall that $\mathbf{r}\cdot\mathbf{n}|_{\partial K} \in \mathcal{R}_k(\partial K)$) and therefore

$$(\nabla u_h, \mathbf{r})_{\mathcal{T}_h} = 0 \qquad \forall \mathbf{r} \in \mathbf{V}_h.$$

Taking $\mathbf{r} = \nabla u_h$, we prove that u_h is piecewise constant, i.e., $u_h|_K \equiv c_K$ for all K. Going again to the boundary, it follows that $\widehat{u}_h|_{\partial K} = c_K$ for all K, but, since $\widehat{u}_h$ is single-valued on interior faces and vanishes on the boundary, this proves that $u_h = 0$ and $\widehat{u}_h = 0$, which completes the proof.

Before we embark on the convergence analysis of this new HDG scheme, let us introduce the associated projection. The equations defining the new projection look much more involved than those in (3.4), but we will next see how they adapt perfectly to the HDG+ method.

The HDG+ projection. Given sufficiently smooth $(\mathbf{q}, u) : K \to \mathbb{R}^d \times \mathbb{R}$, we define

$$(\boldsymbol{\Pi}^{\mathrm{HDG+}}\mathbf{q}, \Pi^{\mathrm{HDG+}}u) := (\boldsymbol{\Pi}_q^{\mathrm{HDG+}}(\mathbf{q}, u), \Pi_u^{\mathrm{HDG+}}(\mathbf{q}, u))$$

as the elements

$$(\mathbf{q}_K, u_K) \in \boldsymbol{\mathcal{P}}_k(K) \times \mathcal{P}_{k+1}(K)$$

that solve the equations

$$(\mathbf{q}_K, \mathbf{r})_K = (\mathbf{q}, \mathbf{r})_K \qquad \forall \mathbf{r} \in \boldsymbol{\mathcal{P}}_{k-1}(K), \tag{4.5a}$$

$$(u_K, v)_K = (u, v)_K \qquad \forall v \in \mathcal{P}_{k-1}(K), \tag{4.5b}$$

$$\langle \mathbf{q}_K \cdot \mathbf{n} + \tau \mathrm{P} u_K, \mu \rangle_{\partial K} = \langle \mathbf{q} \cdot \mathbf{n} + \tau u, \mu \rangle_{\partial K} \qquad \forall \mu \in \mathcal{R}_k(\partial K), \tag{4.5c}$$

$$(\operatorname{div} \mathbf{q}_K, w)_K + \langle \tau \mathrm{P} u_K, w \rangle_{\partial K} = (\operatorname{div} \mathbf{q}, w)_K + \langle \tau \mathrm{P} u, w \rangle_{\partial K} \qquad \forall w \in \widetilde{\mathcal{P}}_{k+1}(K). \tag{4.5d}$$

Some comments on the projection. Using integration by parts, it is clear that (4.5d) is equivalent to

$$(\mathbf{q}_K, \nabla w)_K - \langle \mathbf{q}_K \cdot \mathbf{n} + \tau \mathrm{P} u_K, w \rangle_{\partial K} = (\mathbf{q}, \nabla w)_K - \langle \mathbf{q} \cdot \mathbf{n} + \tau \mathrm{P} u, w \rangle_{\partial K} \quad \forall w \in \widetilde{\mathcal{P}}_{k+1}(K).$$

However, by (4.5a) and (4.5c), it is clear that the projection also satisfies

$$(\mathbf{q}_K, \nabla w)_K - \langle \mathbf{q}_K \cdot \mathbf{n} + \tau \mathrm{P} u_K, w \rangle_{\partial K} = (\mathbf{q}, \nabla w)_K - \langle \mathbf{q} \cdot \mathbf{n} + \tau \mathrm{P} u, w \rangle_{\partial K} \quad \forall w \in \mathcal{P}_{k+1}(K),$$

and therefore (4.5d) can be substituted by

$$(\operatorname{div} \mathbf{q}_K, w)_K + \langle \tau \mathrm{P} u_K, w \rangle_{\partial K} = (\operatorname{div} \mathbf{q}, w)_K + \langle \tau \mathrm{P} u, w \rangle_{\partial K} \quad \forall w \in \mathcal{P}_{k+1}(K). \tag{4.6}$$

This means that **weak commutativity** (compare with (3.6)) is now an explicit requirement in the definition of the projection. Also note that in the right-hand side of (4.5c) we can replace u by $\mathrm{P}u$. As a preview of the analysis to come, let us mention here that the projection is defined so that if $(\mathbf{q}, u)$ is the solution to the model problem (2.44), and $(\boldsymbol{\Pi}\mathbf{q}, \Pi u)$ is its HDG+ projection, then

$$
\begin{aligned}
(\kappa^{-1}\boldsymbol{\Pi}\mathbf{q}, \mathbf{r})_{\mathscr{T}_h} - (\Pi u, \operatorname{div}\mathbf{r})_{\mathscr{T}_h} + \langle \mathrm{P}u, \mathbf{r}\cdot\mathbf{n}\rangle_{\partial\mathscr{T}_h} &= (\kappa^{-1}(\boldsymbol{\Pi}\mathbf{q}-\mathbf{q}), \mathbf{r})_{\mathscr{T}_h}, &(4.7a)\\
(\operatorname{div}\boldsymbol{\Pi}\mathbf{q}, w)_{\mathscr{T}_h} + \langle \tau(\mathrm{P}\Pi u - \mathrm{P}u), w\rangle_{\partial\mathscr{T}_h} &= (f, w)_{\mathscr{T}_h}, &(4.7b)\\
\langle \boldsymbol{\Pi}\mathbf{q}\cdot\mathbf{n} + \tau(\mathrm{P}\Pi u - \mathrm{P}u), \mu_1\rangle_{\partial\mathscr{T}_h\setminus\Gamma} &= 0, &(4.7c)\\
\langle \mathrm{P}u, \mu_2\rangle_\Gamma &= \langle g, \mu_2\rangle_\Gamma, &(4.7d)
\end{aligned}
$$

for all $\mathbf{r}\in\mathbf{V}_h$, $w\in W_h$, $\mu_1\in M_h^\circ$ and $\mu_2\in M_h^\Gamma$, which will be the trigger for our projection-based analysis.

4.2 Analysis of the HDG+ Projection

In this section, we will try to mimic, as much as possible, the analysis of the HDG projection of Sect. 3.3, by a change to the reference domain. While most of the analysis will be relatively similar, there are some new techniques required in this proof. In particular, we will have to be careful with scaling properties for the stabilization parameter. We will need to keep track of the two components of the projection simultaneously, which will force us to momentarily abandon the abuse of notation $(\boldsymbol{\Pi}\mathbf{q}, \Pi u)$ used to denote the projection. We will also work with some more general stabilization parameters for some of the arguments. Given $\tau\in\mathscr{R}_0(\partial K)$ and a pair of functions $(\mathbf{q}, u)$, we will denote

$$
(\mathbf{q}_K, u_K) = \Pi(\mathbf{q}, u; \tau) \in \boldsymbol{\mathscr{P}}_k(K)\times\mathscr{P}_{k+1}(K)
$$

to the solution of the equations

$$
\begin{aligned}
(\mathbf{q}_K, \mathbf{r})_K &= (\mathbf{q}, \mathbf{r})_K & \forall \mathbf{r}\in\boldsymbol{\mathscr{P}}_{k-1}(K), \quad (4.8a)\\
(u_K, v)_K &= (u, v)_K & \forall v\in\mathscr{P}_{k-1}(K), \quad (4.8b)\\
\langle \mathbf{q}_K\cdot\mathbf{n} + \tau\mathrm{P}u_K, \mu\rangle_{\partial K} &= \langle \mathbf{q}\cdot\mathbf{n} + \tau u, \mu\rangle_{\partial K} & \forall \mu\in\mathscr{R}_k(\partial K), \quad (4.8c)\\
(\operatorname{div}\mathbf{q}_K, v)_K + \langle \tau\mathrm{P}u_K, v\rangle_{\partial K} &= (\operatorname{div}\mathbf{q}, v)_K + \langle \tau\mathrm{P}u, v\rangle_{\partial K} & \forall v\in\mathscr{P}_{k+1}(K). \quad (4.8d)
\end{aligned}
$$

As we have already explained, the test in (4.8d) can be changed to the space of homogeneous polynomials $\widetilde{\mathscr{P}}_{k+1}(K)$, since the equations for $v\in\mathscr{P}_k(K)$ are already covered by (4.8a) and (4.8c).

Proposition 4.2 (The projection is well defined) *For any $\tau\in\mathscr{R}_0(\partial K)$ such that $\tau > 0$ or $\tau < 0$, Eqs.* (4.8) *are uniquely solvable.*

Proof Let us start with a count of the number of equations. Note that Eqs. (4.8) (eliminating redundant equations in (4.8d)) amount to a system of

$$
\begin{aligned}
N_{\mathrm{eq}} &:= d \dim \mathcal{P}_{k-1}(K) + \dim \mathcal{P}_{k-1}(K) \\
&\quad + (d+1)\dim P_k(e) + \dim \widetilde{\mathcal{P}}_{k+1}(K) \\
&= d \dim \mathcal{P}_k(K) + \dim \mathcal{P}_k(K) + \dim \widetilde{\mathcal{P}}_{k+1}(K) \\
&= \dim \boldsymbol{\mathcal{P}}_k(K) + \dim \mathcal{P}_{k+1}(K),
\end{aligned}
$$

which shows that the number of equations and unknowns are the same. We therefore only need to show uniqueness of solution for the homogeneous system. Let then $(\mathbf{q}_K, u_K) \in \boldsymbol{\mathcal{P}}_k(K) \times \mathcal{P}_{k+1}(K)$ satisfy

$$
\begin{aligned}
(\mathbf{q}_K, \mathbf{r})_K &= 0 & \forall \mathbf{r} &\in \boldsymbol{\mathcal{P}}_{k-1}(K), & \text{(4.9a)}\\
(u_K, v)_K &= 0 & \forall v &\in \mathcal{P}_{k-1}(K), & \text{(4.9b)}\\
\langle \mathbf{q}_K \cdot \mathbf{n} + \tau u_K, \mu\rangle_{\partial K} &= 0 & \forall \mu &\in \mathcal{R}_k(\partial K), & \text{(4.9c)}\\
(\operatorname{div} \mathbf{q}_K, v)_K + \langle \tau \mathrm{P} u_K, v\rangle_{\partial K} &= 0 & \forall v &\in \mathcal{P}_{k+1}(K). & \text{(4.9d)}
\end{aligned}
$$

Noting that $\operatorname{div} \mathbf{q}_K \in \mathcal{P}_{k-1}(K)$, it follows from (4.9b) and (4.9d) that

$$\langle \tau \mathrm{P} u_K, u_K\rangle_{\partial K} = 0.$$

Since τ is either strictly positive or strictly negative and piecewise constant, this shows that $\mathrm{P} u_K = 0$. Taking now $\mu = \mathbf{q}_K \cdot \mathbf{n}$ in (4.9c), it follows that $\mathbf{q}_K \cdot \mathbf{n} = 0$ and at the same time $\mathbf{q}_K \in \boldsymbol{\mathcal{P}}_k^{\perp}(K)$. By Lemma 2.1(b), this shows that $\mathbf{q}_K = \mathbf{0}$. Finally

$$
\begin{aligned}
0 &= \langle u_K, \nabla u_K \cdot \mathbf{n}\rangle_{\partial K} & &(\mu = \tau^{-1}\nabla u_K \cdot \mathbf{n} \text{ in (4.9c)})\\
&= (\Delta u_K, u_K)_K + (\nabla u_K, \nabla u_K)_K & &\\
&= \|\nabla u_K\|_K^2, & &(v = \Delta u_K \text{ in (4.9b)})
\end{aligned}
$$

which proves that u_K is constant. Since $\mathrm{P} u_K = 0$, this finally proves that $u_K = 0$, which completes the proof.

In order to make a change to the reference element, we need a small lemma that connects bilinear forms on the boundary of the element with bilinear forms on the boundary of the reference element.

Lemma 4.1 *For all $u \in H^1(K)$ and $\mu \in L^2(\partial K)$,*

$$\langle \tau \mathrm{P} u, \mu\rangle_{\partial K} = \langle \check{\tau}\, \widehat{\mathrm{P}} \widehat{u}, \widehat{\mu}\rangle_{\partial \widehat{K}},$$

where $\widehat{\mathrm{P}} : L^2(\partial \widehat{K}) \to \mathcal{R}_k(\partial \widehat{K})$ is the orthogonal projection onto $\mathcal{R}_k(\partial \widehat{K})$.

Proof Note first that

$$\mathcal{R}_k(\partial K) \ni \mu \longmapsto \check{\mu} \in \mathcal{R}_k(\partial \widehat{K})$$

is a bijection. Since $\widehat{\mathrm{P}u} \in \mathscr{R}_k(\partial \widehat{K})$ and for all $\mu \in \mathscr{R}_k(\partial K)$,

$$\begin{aligned}
\langle \widehat{\mathrm{P}u}, \check{\mu}\rangle_{\partial \widehat{K}} &= \langle \mathrm{P}u, \mu\rangle_{\partial K} && \text{(by (1.1c))}\\
&= \langle u, \mu\rangle_{\partial K}\\
&= \langle \widehat{u}, \check{\mu}\rangle_{\partial \widehat{K}}, && \text{(by (1.1c) and (1.2))}
\end{aligned}$$

it follows that

$$\widehat{\mathrm{P}u} = \widehat{\mathrm{P}}\widehat{u}. \tag{4.10}$$

Finally, for all $\mu \in L^2(\partial K)$

$$\begin{aligned}
\langle \tau \mathrm{P}u, \mu\rangle_{\partial K} &= \langle \widetilde{\tau \mathrm{P}u}, \widehat{\mu}\rangle_{\partial \widehat{K}} && \text{(by (1.1c))}\\
&= \langle \check{\tau}\, \widehat{\mathrm{P}u}, \widehat{\mu}\rangle_{\partial \widehat{K}} && \text{(easy computation)}\\
&= \langle \check{\tau}\, \widehat{\mathrm{P}}\widehat{u}, \widehat{\mu}\rangle_{\partial \widehat{K}}, && \text{(by (4.10))}
\end{aligned}$$

and the proof is finished.

Proposition 4.3 (Change to the reference element) *Let* $(\mathbf{q}, u) : K \to \mathbb{R}^d \times \mathbb{R}$ *be sufficiently smooth and let* $\tau \in \mathscr{R}_0(\partial K)$ *be either strictly positive or strictly negative. If* $(\mathbf{q}_K, u_K) = \Pi(\mathbf{q}, u; \tau)$*, then*

$$(\widehat{\mathbf{q}_K}, \widehat{u_K}) = \widehat{\Pi}(\widehat{\mathbf{q}}, \widehat{u}; \check{\tau}).$$

Proof The result is a simple consequence of the changes of variables in Sect. 1.1 enhanced with Lemma 4.1. Using Lemma 4.1, (1.1), (1.4a), and (1.4c), it follows that

$$\begin{aligned}
(\mathbf{q}, \mathbf{r})_K &= (\widehat{\mathbf{q}}, \check{\mathbf{r}})_{\widehat{K}},\\
(u, v)_K &= (\widehat{u}, \check{v})_{\widehat{K}},\\
\langle \mathbf{q}\cdot\mathbf{n} + \tau \mathrm{P}u, \mu\rangle_{\partial K} &= \langle \widehat{\mathbf{q}}\cdot\widehat{\mathbf{n}} + \check{\tau}\widehat{\mathrm{P}}\widehat{u}, \widehat{\mu}\rangle_{\partial \widehat{K}},\\
(\operatorname{div}\mathbf{q}, v)_K + \langle \tau\, \mathrm{P}u, v\rangle_{\partial K} &= (\widehat{\operatorname{div}}\,\widehat{\mathbf{q}}, \widehat{v})_{\widehat{K}} + \langle \check{\tau}\widehat{\mathrm{P}}\widehat{u}, \widehat{v}\rangle_{\partial \widehat{K}},
\end{aligned}$$

for generic $\mathbf{q}$, $\mathbf{r}$, u, and v. At the same time, all the following transformations

$$\begin{aligned}
\boldsymbol{\mathscr{P}}_{k-1}(K) \ni \mathbf{r} \quad &\longmapsto \quad \check{\mathbf{r}} \in \boldsymbol{\mathscr{P}}_{k-1}(\widehat{K}),\\
\mathscr{P}_{k-1}(K) \ni v \quad &\longmapsto \quad \check{v} \in \mathscr{P}_{k-1}(\widehat{K}),\\
\mathscr{R}_k(\partial K) \ni \mu \quad &\longmapsto \quad \widehat{\mu} \in \mathscr{R}_k(\partial \widehat{K}),\\
\mathscr{P}_{k+1}(K) \ni v \quad &\longmapsto \quad \widehat{v} \in \mathscr{P}_{k+1}(\widehat{K}),
\end{aligned}$$

are bijections. The proof is then straightforward.

Proposition 4.4 (Stability in the reference domain) *Let* $0 < \widehat{\tau}_{\min} \leqslant \widehat{\tau}_{\max}$. *There exists* $\widehat{C} = \widehat{C}(\widehat{\tau}_{\min}, \widehat{\tau}_{\max}, k)$ *such that*

$$\|\widehat{\Pi}(\mathbf{q}, u; \widehat{\tau})\|_{\widehat{K}} \leqslant \widehat{C}\|(\mathbf{q}, u)\|_{1,\widehat{K}} \qquad \forall(\mathbf{q}, u) \in H^1(\widehat{K}; \mathbb{R}^{d+1})$$

and for all

$$\widehat{\tau} \in \mathscr{R}_0(\partial\widehat{K}) \qquad \widehat{\tau}_{\min} \leqslant \widehat{\tau} \leqslant \widehat{\tau}_{\max} \quad on\ \partial\widehat{K}. \tag{4.11}$$

Proof Upon numbering the faces of ∂K, we can identify

$$\widehat{\tau} \quad \longleftrightarrow \quad \mathbf{x} := (x_1, \ldots, x_{d+1}) \in [\widehat{\tau}_{\min}, \widehat{\tau}_{\max}]^{d+1} \subset \mathbb{R}^{d+1}_+.$$

We can then define bounded bilinear maps

$$a_j : H^1(\widehat{K}; \mathbb{R}^{d+1}) \times H^1(\widehat{K}; \mathbb{R}^{d+1}) \to \mathbb{R}$$

such that Eqs. (4.8) (written in the reference element $\widehat{K}$ and with stabilization parameter $\widehat{\tau}$) are equivalent to

$$\widehat{\Pi}(\mathbf{q}, u; \widehat{\tau}) \in \boldsymbol{\mathscr{P}}_k(\widehat{K}) \times \mathscr{P}_{k+1}(\widehat{K}), \tag{4.12a}$$

$$a_0(\widehat{\Pi}(\mathbf{q}, u; \widehat{\tau}) - (\mathbf{q}, u), (\mathbf{r}, v, \mu)) + \sum_{j=1}^{d+1} x_j a_j(\widehat{\Pi}(\mathbf{q}, u; \widehat{\tau}) - (\mathbf{q}, u), (\mathbf{r}, v, \mu)) = 0$$
$$\forall(\mathbf{r}, v, \mu) \in \boldsymbol{\mathscr{P}}_{k-1}(\widehat{K}) \times (\mathscr{P}_{k-1}(\widehat{K}) \oplus \widetilde{\mathscr{P}}_{k+1}(\widehat{K})) \times \mathscr{R}_k(\partial\widehat{K}). \tag{4.12b}$$

Equations (4.12) are uniquely solvable for all $\mathbf{x} \in \mathbb{R}^{d+1}_+$ (because of Proposition 4.2) and thus define a bounded operator

$$T(\mathbf{x}) \in \mathscr{B}(H^1(\widehat{K}; \mathbb{R}^{d+1}); L^2(\widehat{K}; \mathbb{R}^{d+1})).$$

The function

$$T : \mathbb{R}^{d+1}_+ \longrightarrow \mathscr{B}(H^1(\widehat{K}; \mathbb{R}^{d+1}); L^2(\widehat{K}; \mathbb{R}^{d+1}))$$

is a rational function, and therefore it is a continuous function in $\mathbb{R}^{d+1}_+$. It is therefore bounded on the compact set $[\widehat{\tau}_{\min}, \widehat{\tau}_{\max}]^{d+1}$. The quantity $\widehat{C}$ in the statement is an upper bound of T on this compact set.

Proposition 4.5 (Reference-domain estimate) *Let* $\widehat{\tau}$ *satisfy* (4.11) *and* $0 \neq \eta \in \mathbb{R}$. *If*

$$(\widehat{\mathbf{q}}_h, \widehat{u}_h) = \widehat{\Pi}(\widehat{\mathbf{q}}, \widehat{u}; \eta\widehat{\tau})$$

and $(\widehat{\mathbf{q}}, \widehat{u}) \in \mathbf{H}^m(\widehat{K}) \times H^{m+1}(\widehat{K})$ *with* $1 \leqslant m \leqslant k+1$, *then*

$$\|\widehat{\mathbf{q}}_h - \widehat{\mathbf{q}}\|_{\widehat{K}} + |\eta|\, \|\widehat{u}_h - \widehat{u}\|_{\widehat{K}} \leqslant \widehat{D}(|\widehat{\mathbf{q}}|_{m,\widehat{K}} + |\eta|\, |\widehat{u}|_{m+1,\widehat{K}}),$$

where the constant $\widehat{D}$ *depends on* $\widehat{C}$ *of Proposition 4.4 and on* k.

Proof The result follows from a simple scaling argument in the stabilization parameter together with traditional approximation inequalities. It is simple to see that

$$(\widehat{\mathbf{q}}_h, \eta \widehat{u}_h) = \widehat{\Pi}(\widehat{\mathbf{q}}, \eta \widehat{u}; \widehat{\tau}),$$

and we can therefore use Proposition 4.4 to estimate

$$\|\widehat{\mathbf{q}}_h\|_{\widehat{K}} + |\eta|\, \|\widehat{u}_h\|_{\widehat{K}} \leqslant \widehat{C}(\|\widehat{\mathbf{q}}\|_{1,\widehat{K}} + |\eta|\, \|\widehat{u}\|_{1,\widehat{K}}).$$

If $\mathbf{p} \in \mathscr{P}_k(\widehat{K})$ is the best $\mathbf{H}^1(\widehat{K})$ approximation of $\widehat{\mathbf{q}}$ and $v \in \mathscr{P}_{k+1}(\widehat{K})$ is the best $H^1(\widehat{K})$ approximation of $\widehat{u}$, using the fact that $(\mathbf{p}, \eta v) = \widehat{\Pi}(\mathbf{p}, \eta v; \widehat{\tau})$ and $\widehat{\Pi}$ is linear, we obtain the estimate

$$\|\widehat{\mathbf{q}}_h - \widehat{\mathbf{q}}\|_{\widehat{K}} + |\eta|\, \|\widehat{u}_h - \widehat{u}\|_{\widehat{K}} \leqslant (1 + \widehat{C})(\|\widehat{\mathbf{q}} - \mathbf{p}\|_{1,\widehat{K}} + |\eta|\, \|\widehat{u} - v\|_{1,\widehat{K}}),$$

or, equivalently,

$$\begin{aligned}&\|\widehat{\mathbf{q}}_h - \widehat{\mathbf{q}}\|_{\widehat{K}} + |\eta|\, \|\widehat{u}_h - \widehat{u}\|_{\widehat{K}}\\ &\leqslant (1 + \widehat{C})\left(\inf_{\mathbf{p} \in \mathscr{P}_k(\widehat{K})} \|\widehat{\mathbf{q}} - \mathbf{p}\|_{1,\widehat{K}} + |\eta| \inf_{v \in \mathscr{P}_{k+1}(\widehat{K})} \|\widehat{u} - v\|_{1,\widehat{K}} \right).\end{aligned}$$

The result follows now from a compactness (Rellich–Kondrachov, Bramble–Hilbert, or Deny–Lions) argument.

Proposition 4.6 (Estimates for the HDG+ projection) *Let*

$$(\mathbf{q}_K, u_K) = \Pi^{\mathrm{HDG+}}(\mathbf{q}, u; \tau),$$

where

$$\tau \in \mathscr{R}_0(\partial K), \qquad \tau > 0 \ \ or \ \ \tau < 0, \qquad |\tau| \approx h_K^{-1}.$$

If $(\mathbf{q}, u) \in \mathbf{H}^m(K) \times H^{m+1}(K)$ *with* $1 \leqslant m \leqslant k+1$, *then*

$$h_K \|\mathbf{q} - \mathbf{q}_K\|_K + \|u - u_K\|_K \lesssim h_K^{m+1}(|\mathbf{q}|_{m,K} + |u|_{m+1,K}).$$

Proof By Proposition 4.3 (change to the reference element)

$$(\widehat{\mathbf{q}_K}, \widehat{u_K}) = \widehat{\Pi}(\widehat{\mathbf{q}}, \widehat{u}; \check{\tau}) = \widehat{\Pi}(\widehat{\mathbf{q}}, \widehat{u}; \pm h_K^{d-2} \widehat{\tau}),$$

where by hypothesis on τ and (1.6) (the scaling property for the transformation $\tau \mapsto \check{\tau}$), we have $\widehat{\tau} \approx 1$ and the sign is chosen depending on the sign of τ so that $\widehat{\tau} > 0$. Therefore

$$
\begin{aligned}
&\|\mathbf{q}-\mathbf{q}_K\|_K + h_K^{-1}\|u-u_K\|_K \\
&\approx h_K^{1-\frac{d}{2}}\|\widehat{\mathbf{q}}-\widehat{\mathbf{q}_K}\|_{\widehat{K}} + h_K^{\frac{d}{2}-1}\|\widehat{u}-\widehat{u_K}\|_{\widehat{K}} && \text{(by (1.9))}\\
&= h_K^{1-\frac{d}{2}}(\|\widehat{\mathbf{q}}-\widehat{\mathbf{q}_K}\|_{\widehat{K}} + h_K^{d-2}\|\widehat{u}-\widehat{u_K}\|_{\widehat{K}})\\
&\lesssim h_K^{1-\frac{d}{2}}(|\widehat{\mathbf{q}}|_{m,\widehat{K}} + h_K^{d-2}|\widehat{u}|_{m+1,\widehat{K}}) && \text{(Proposition 4.5)}\\
&= h_K^{1-\frac{d}{2}}|\widehat{\mathbf{q}}|_{m,\widehat{K}} + h_K^{\frac{d}{2}-1}|\widehat{u}|_{m+1,\widehat{K}}\\
&\approx h_K^m|\mathbf{q}|_{m,K} + h_K^m|u|_{m+1,K}, && \text{(by (1.9))}
\end{aligned}
$$

which completes the proof.

4.3 Analysis of the HDG+ Method

Thanks to the existence of a tailored projection, the analysis of the HDG+ method is practically identical (up to minor details that need to be adjusted) to the analysis of the HDG method given in Sect. 3.4. We thus consider the solution $(\mathbf{q}, u)$ of the model Eqs. (2.14), its HDG+ projection $(\boldsymbol{\Pi}\mathbf{q}, \Pi u) = \Pi^{\mathrm{HDG+}}(\mathbf{q}, u; \tau)$ defined element by element by (4.5), and the associated discrete solution given by Eqs. (4.3). As usual, we consider the errors

$$\boldsymbol{\varepsilon}_h^q := \boldsymbol{\Pi}\mathbf{q} - \mathbf{q}_h \in \mathbf{V}_h, \qquad \varepsilon_h^u := \Pi u - u_h \in W_h, \qquad \widehat{\varepsilon}_h^u := \mathrm{P}u - \widehat{u}_h \in M_h.$$

Subtracting the discrete Eqs. (4.3) from the equations satisfied by the projections (4.7), we arrive at the **error equations** (compare with (3.18) and note that we have changed the sign in the third equation for convenience)

$$
\begin{aligned}
&(\kappa^{-1}\boldsymbol{\varepsilon}_h^q, \mathbf{r})_{\mathscr{T}_h} - (\varepsilon_h^u, \operatorname{div}\mathbf{r})_{\mathscr{T}_h} + \langle\widehat{\varepsilon}_h^u, \mathbf{r}\cdot\mathbf{n}\rangle_{\partial\mathscr{T}_h} = (\kappa^{-1}(\boldsymbol{\Pi}\mathbf{q}-\mathbf{q}), \mathbf{r})_{\mathscr{T}_h}, && \text{(4.13a)}\\
&(\operatorname{div}\boldsymbol{\varepsilon}_h^q, w)_{\mathscr{T}_h} + \langle\tau(\mathrm{P}\varepsilon_h^u - \widehat{\varepsilon}_h^u), w\rangle_{\partial\mathscr{T}_h} = 0, && \text{(4.13b)}\\
&-\langle\boldsymbol{\varepsilon}_h^q\cdot\mathbf{n} + \tau(\mathrm{P}\varepsilon_h^u - \widehat{\varepsilon}_h^u), \mu_1\rangle_{\partial\mathscr{T}_h\setminus\Gamma} = 0, && \text{(4.13c)}\\
&\langle\widehat{\varepsilon}_h^u, \mu_2\rangle_\Gamma = 0, && \text{(4.13d)}
\end{aligned}
$$

for all $\mathbf{r} \in \mathbf{V}_h$, $w \in W_h$, $\mu_1 \in M_h^\circ$ and $\mu_2 \in M_h^\Gamma$. Testing the first three equations of (4.13) with $\mathbf{r} = \boldsymbol{\varepsilon}_h^q$, $w = \varepsilon_h^u$, $\mu = \widehat{\varepsilon}_h^u$, adding the results, and noticing that $\widehat{\varepsilon}_h^u = 0$ on Γ, we reach the **energy identity**

$$(\kappa^{-1}\boldsymbol{\varepsilon}_h^q, \boldsymbol{\varepsilon}_h^q)_{\mathscr{T}_h} + \langle\tau(\mathrm{P}\varepsilon_h^u - \widehat{\varepsilon}_h^u), \mathrm{P}\varepsilon_h^u - \widehat{\varepsilon}_h^u\rangle_{\partial\mathscr{T}_h} = (\kappa^{-1}(\boldsymbol{\Pi}\mathbf{q}-\mathbf{q}), \boldsymbol{\varepsilon}_h^q)_{\mathscr{T}_h}, \qquad (4.14)$$

from which we obtain the first error estimate

$$\|\boldsymbol{\varepsilon}_h^q\|_{\kappa^{-1}}^2 + |\mathrm{P}\varepsilon_h^u - \widehat{\varepsilon}_h^u|_\tau^2 \leqslant \|\boldsymbol{\Pi}\mathbf{q} - \mathbf{q}\|_{\kappa^{-1}}^2. \qquad (4.15)$$

(We will still use the symbol $|\cdot|_\tau$, although with $\tau \neq 0$, this is a norm.) The estimate for the flux

$$\widehat{\varepsilon}_h^q := \mathrm{P}(\mathbf{q}\cdot\mathbf{n}) - \widehat{\mathbf{q}}_h\cdot\mathbf{n} = \boldsymbol{\varepsilon}_h^q\cdot\mathbf{n} + \tau(\mathrm{P}\varepsilon_h^u - \widehat{\varepsilon}_h^u)$$

is now easy:

$$\begin{aligned}
\|\widehat{\varepsilon}_h^q\|_h &\leqslant \|\boldsymbol{\varepsilon}_h^q\cdot\mathbf{n}\|_h + \|\tau(\mathrm{P}\varepsilon_h^u - \widehat{\varepsilon}_h^u)\|_h \\
&\lesssim \|\boldsymbol{\varepsilon}_h^q\|_\Omega + |\mathrm{P}\varepsilon_h^u - \widehat{\varepsilon}_h^u|_\tau && \text{(scaling and } \tau\approx h_K^{-1}) \\
&\lesssim \|\boldsymbol{\Pi}\mathbf{q} - \mathbf{q}\|_\Omega. && \text{(by (4.15))}
\end{aligned}$$

Note that

$$\|\mu\|_h \lesssim h|\mu|_\tau, \tag{4.16}$$

since $\tau \approx h_K^{-1}$ and, while P is the orthogonal projection onto $\mathscr{R}_k(\partial K)$ using the $L^2(\partial K)$ inner product, it is also the best approximation operator with respect to the scaled norm $\|\cdot\|_h$, which is due to the fact that approximation on $\mathscr{R}_k(\partial K)$ is carried out separately on each face and therefore scaling does not affect it. This implies that

$$\begin{aligned}
\|\widehat{\varepsilon}_h^u\|_h &\leqslant \|\mathrm{P}\varepsilon_h^u\|_h + \|\mathrm{P}\varepsilon_h^u - \widehat{\varepsilon}_h^u\|_h \\
&\lesssim \|\varepsilon_h^u\|_h + h|\mathrm{P}\varepsilon_h^u - \widehat{\varepsilon}_h^u|_\tau && \text{(P is best approx and (4.16))} \\
&\lesssim \|\varepsilon_h^u\|_\Omega + h\|\boldsymbol{\Pi}\mathbf{q} - \mathbf{q}\|_\Omega, && \text{(scaling argument and (4.15))}
\end{aligned}$$

where in the last inequality we have used a discrete trace inequality followed by an inverse inequality. Pending an estimate for $\|\varepsilon_h^u\|_\Omega$, which we will obtain with duality arguments, this gives superconvergence for $\|\widehat{\varepsilon}_h^u\|_h$. As opposed to the HDG method of Sect. 3.1, we do not need to use the BDM projection for the estimate on $\|\widehat{\varepsilon}_h^u\|_h$, which makes the argument work for $k = 0$ as well.

To derive estimates on ε_h^u, we use a **duality argument**. We consider the dual problem

$$\begin{aligned}
\kappa^{-1}\boldsymbol{\xi} - \nabla\theta &= \mathbf{0} && \text{in } \Omega, && (4.17\text{a}) \\
-\operatorname{div}\boldsymbol{\xi} &= \varepsilon_h^u && \text{in } \Omega, && (4.17\text{b}) \\
\theta &= 0 && \text{on } \Gamma, && (4.17\text{c})
\end{aligned}$$

and its projection $(\boldsymbol{\Pi}\boldsymbol{\xi}, \Pi\theta) := \Pi^{\mathrm{HDG+}}(\boldsymbol{\xi}, \theta; -\tau)$ which satisfies the equations

$$\begin{aligned}
(\kappa^{-1}\boldsymbol{\Pi}\boldsymbol{\xi}, \mathbf{r})_{\mathscr{T}_h} + (\Pi\theta, \operatorname{div}\mathbf{r})_{\mathscr{T}_h} - \langle \mathrm{P}\theta, \mathbf{r}\cdot\mathbf{n}\rangle_{\partial\mathscr{T}_h} &= (\kappa^{-1}(\boldsymbol{\Pi}\boldsymbol{\xi} - \boldsymbol{\xi}), \mathbf{r})_{\mathscr{T}_h} && \forall \mathbf{r}\in\mathbf{V}_h, \\
-(\operatorname{div}\boldsymbol{\Pi}\boldsymbol{\xi}, w)_{\mathscr{T}_h} + \langle \tau(\mathrm{P}\Pi\theta - \mathrm{P}\theta), w\rangle_{\partial\mathscr{T}_h} &= (\varepsilon_h^u, w)_{\mathscr{T}_h} && \forall w\in W_h, \\
\langle \boldsymbol{\Pi}\boldsymbol{\xi}\cdot\mathbf{n} - \tau(\mathrm{P}\Pi\theta - \mathrm{P}\theta), \mu\rangle_{\partial\mathscr{T}_h\setminus\Gamma} &= 0 && \forall \mu\in M_h^\circ, \\
\langle \mathrm{P}\theta, \mu\rangle_\Gamma &= 0 && \forall \mu \in M_h^\Gamma.
\end{aligned}$$

We can proceed as in Sect. 3.4.2 and reach the identity

$$\|\varepsilon_h^u\|_\Omega^2 = (\boldsymbol{\Pi}\boldsymbol{\xi} - \boldsymbol{\xi}, \kappa^{-1}(\mathbf{q}_h - \mathbf{q}))_{\mathscr{T}_h} + (\nabla\theta, \boldsymbol{\Pi}\mathbf{q} - \mathbf{q})_{\mathscr{T}_h}.$$

For $k \geqslant 1$, we can consider $\theta_h \in \prod_{K\in\mathscr{T}_h} \mathscr{P}_1(K)$ to be the best $\prod_{K\in\mathscr{T}_h} H^1(K)$ approximation of θ to obtain

$$\begin{aligned}
\|\varepsilon_h^u\|_\Omega^2 &= (\boldsymbol{\Pi}\boldsymbol{\xi} - \boldsymbol{\xi}, \kappa^{-1}(\mathbf{q}_h - \mathbf{q}))_{\mathscr{T}_h} + (\nabla\theta - \nabla\theta_h, \boldsymbol{\Pi}\mathbf{q} - \mathbf{q})_{\mathscr{T}_h} && \text{(by (4.5a))}\\
&\lesssim h|\boldsymbol{\xi}|_{1,\Omega}\|\mathbf{q}_h - \mathbf{q}\|_\Omega + h|\theta|_{2,\Omega}\|\boldsymbol{\Pi}\mathbf{q} - \mathbf{q}\|_\Omega && \text{(Prop 4.6 \& poly approx)}\\
&\lesssim h\|\varepsilon_h^u\|_\Omega\|\boldsymbol{\Pi}\mathbf{q} - \mathbf{q}\|_\Omega, && \text{(regularity and (4.15))}
\end{aligned}$$

that is,

$$\|\varepsilon_h^u\|_\Omega \lesssim h\|\boldsymbol{\Pi}\mathbf{q} - \mathbf{q}\|_\Omega. \tag{4.18}$$

Exercises

1. Prove that for all $u \in H^1(K)$ and $\mathbf{q} \in \mathbf{H}^1(K)$

$$\|u\|_{\partial K} \approx h_K^{\frac{d-1}{2}} \|\widehat{u}\|_{\partial\widehat{K}}, \qquad \|\mathbf{q}\cdot\mathbf{n}\|_{\partial K} \approx h_K^{\frac{1-d}{2}} \|\widehat{\mathbf{q}}\cdot\widehat{\mathbf{n}}\|_{\partial\widehat{K}}.$$

2. Let $\widehat{\tau}$ satisfy (4.11) and $0 \neq \eta \in \mathbb{R}$. If $(\widehat{\mathbf{q}}_h, \widehat{u}_h) = \widehat{\Pi}(\widehat{\mathbf{q}}, \widehat{u}; \eta\widehat{\tau})$ and $(\widehat{\mathbf{q}}, \widehat{u}) \in \mathbf{H}^m(\widehat{K}) \times H^{m+1}(\widehat{K})$ with $1 \leqslant m \leqslant k+1$, show that

$$\|\widehat{\mathbf{q}}_h - \widehat{\mathbf{q}}\|_{\partial\widehat{K}} + |\eta|\,\|\widehat{u}_h - \widehat{u}\|_{\partial\widehat{K}} \leqslant \widehat{E}(|\widehat{\mathbf{q}}|_{m,\widehat{K}} + |\eta|\,|\widehat{u}|_{m+1,\widehat{K}}),$$

where the constant $\widehat{E}$ depends on k, $\widehat{\tau}_{\min}$, and $\widehat{\tau}_{\max}$.
3. Let $(\mathbf{q}_K, u_K) = \Pi^{\mathrm{HDG+}}(\mathbf{q}, u; \tau)$, where

$$\tau \in \mathscr{R}_0(\partial K), \qquad \tau > 0 \ \text{ or } \ \tau < 0, \qquad |\tau| \approx h_K^{-1}.$$

Show that if $(\mathbf{q}, u) \in \mathbf{H}^m(K) \times H^{m+1}(K)$ with $1 \leqslant m \leqslant k+1$, then

$$h_K\|\mathbf{q} - \mathbf{q}_K\|_{\partial K} + \|u - u_K\|_{\partial K} \lesssim h_K^{m+\frac{1}{2}}(|\mathbf{q}|_{m,K} + |u|_{m+1,K}).$$

4. Let $(\boldsymbol{\Pi}\mathbf{q}, \Pi u) = \Pi^{\mathrm{HDG+}}(\mathbf{q}, u; \tau)$, with τ as in the previous exercise. Show that

$$\|\boldsymbol{\Pi}\mathbf{q}\cdot\mathbf{n} + \tau(\mathrm{P}\Pi u - \mathrm{P}u) - \mathbf{q}\cdot\mathbf{n}\|_h \lesssim h^m(|\mathbf{q}|_{m,K} + |u|_{m+1,K}), \qquad 1 \leqslant m \leqslant k+1.$$

4.4 An Extended HDG+ Scheme

In this section, we show an extended version of the HDG method where an additional local field is added to the equations. This makes the local solvers computationally more expensive, but the size of the global system is not changed. The reason to include an additional local field is to avoid inverting the diffusion coefficient matrix. In principle, this is not particularly important in linear problems, but this is very relevant in the case of nonlinear material laws. The idea can be traced back to [115, 116], where this extended scheme was applied to elasticity. See also the work of Amiya Pani [120] for nonlinear elliptic problems and [55] for a supercloseness analysis. We will present it here in the context of HDG+ methods, but the same ideas apply to the classical HDG scheme.

Formulation and discretization. We are still going to work on the diffusion equation with Dirichlet boundary conditions

$$-\mathrm{div}\,(\kappa\nabla u) = f \quad \text{in } \Omega, \qquad u = g \quad \text{on } \Gamma,$$

but we now consider two auxiliary unknowns $\boldsymbol{\sigma}, \mathbf{q} : \Omega \to \mathbb{R}^d$, so that the equations are

$$\begin{aligned} \kappa\boldsymbol{\sigma} + \mathbf{q} &= \mathbf{0} && \text{in } \Omega, && (4.19a)\\ -\boldsymbol{\sigma} + \nabla u &= \mathbf{0} && \text{in } \Omega, && (4.19b)\\ \mathrm{div}\,\mathbf{q} &= f && \text{in } \Omega, && (4.19c)\\ u &= g && \text{on } \Gamma. && (4.19d) \end{aligned}$$

The discrete spaces are the same ones of the HDG+ method, namely, those given in (4.1). The stabilization function τ is also the same as in the HDG+ scheme, that is, $\tau \equiv h^{-1}$.

We look for

$$(\boldsymbol{\sigma}_h, \mathbf{q}_h, u_h, \widehat{u}_h) \in \mathbf{V}_h \times \mathbf{V}_h \times W_h \times M_h, \tag{4.20a}$$

satisfying

$$\begin{aligned} &(\kappa\boldsymbol{\sigma}_h, \boldsymbol{\eta})_{\mathcal{T}_h} + (\mathbf{q}_h, \boldsymbol{\eta})_{\mathcal{T}_h} &&= 0 && \forall \boldsymbol{\eta} \in \mathbf{V}_h, && (4.20b)\\ &-(\boldsymbol{\sigma}_h, \mathbf{r})_{\mathcal{T}_h} - (u_h, \mathrm{div}\,\mathbf{r})_{\mathcal{T}_h} + \langle \widehat{u}_h, \mathbf{r}\cdot\mathbf{n}\rangle_{\partial\mathcal{T}_h} &&= 0 && \forall \mathbf{r} \in \mathbf{V}_h, && (4.20c)\\ &(\mathrm{div}\,\mathbf{q}_h, w)_{\mathcal{T}_h} + \langle \tau(\mathrm{P}u_h - \widehat{u}_h), w\rangle_{\partial\mathcal{T}_h} &&= (f, w)_{\mathcal{T}_h} && \forall w \in W_h, && (4.20d)\\ &\langle \mathbf{q}_h\cdot\mathbf{n} + \tau(u_h - \widehat{u}_h), \mu\rangle_{\partial\mathcal{T}_h\setminus\Gamma} &&= 0 && \forall \mu \in M_h^\circ, && (4.20e)\\ &\langle \widehat{u}_h, \mu\rangle_\Gamma &&= \langle g, \mu\rangle_\Gamma && \forall \mu \in M_h^\Gamma. && (4.20f) \end{aligned}$$

The proof for unique solvability of (4.20) is left as an exercise.

Energy estimates. The first part of the analysis is quite similar to that of the HDG and HDG+ methods. We need to introduce the orthogonal projection $\mathrm{Q} : L^2(\Omega) \to \mathbf{V}_h$ to control the error of $\boldsymbol{\sigma}$. The other variables are handled as usual, with the HDG+ projection and with the $L^2(\partial K)$ orthogonal projection. We will write the error equations in terms of discrete errors (comparison of numerical solution with the projection)

$$\boldsymbol{\varepsilon}_h^\sigma := \mathrm{Q}\boldsymbol{\sigma} - \boldsymbol{\sigma}_h, \qquad \boldsymbol{\varepsilon}_h^q := \boldsymbol{\Pi}\mathbf{q} - \mathbf{q}_h, \qquad \varepsilon_h^u := \Pi u - u_h, \qquad \widehat{\varepsilon}_h^u := \mathrm{P}u - \widehat{u}_h,$$

and approximation errors

$$\mathbf{e}_h^\sigma := \mathrm{Q}\boldsymbol{\sigma} - \boldsymbol{\sigma}, \quad \mathbf{e}_h^q := \boldsymbol{\Pi}\mathbf{q} - \mathbf{q}.$$

It is simple to see that the error quantities satisfy the discrete error equations

$$
\begin{aligned}
&(\kappa\boldsymbol{\varepsilon}_h^\sigma, \boldsymbol{\eta})_{\mathscr{T}_h} + (\boldsymbol{\varepsilon}_h^q, \boldsymbol{\eta})_{\mathscr{T}_h} &&= (\kappa\mathbf{e}_h^\sigma, \boldsymbol{\eta})_{\mathscr{T}_h} + (\mathbf{e}_h^q, \boldsymbol{\eta})_{\mathscr{T}_h}, && (4.21a)\\
&-(\boldsymbol{\varepsilon}_h^\sigma, \mathbf{r})_{\mathscr{T}_h} - (\varepsilon_h^u, \operatorname{div}\mathbf{r})_{\mathscr{T}_h} + \langle\widehat{\varepsilon}_h^u, \mathbf{r}\cdot\mathbf{n}\rangle_{\partial\mathscr{T}_h} &&= 0, && (4.21b)\\
&(\operatorname{div}\boldsymbol{\varepsilon}_h^q, w)_{\mathscr{T}_h} + \langle\tau(\mathrm{P}\varepsilon_h^u - \widehat{\varepsilon}_h^u), w\rangle_{\partial\mathscr{T}_h} &&= 0, && (4.21c)\\
&-\langle\boldsymbol{\varepsilon}_h^q\cdot\mathbf{n} + \tau(\varepsilon_h^u - \widehat{\varepsilon}_h^u), \mu_1\rangle_{\partial\mathscr{T}_h\setminus\Gamma} &&= 0, && (4.21d)\\
&\langle\widehat{\varepsilon}_h^u, \mu_2\rangle_\Gamma &&= 0, && (4.21e)
\end{aligned}
$$

for all $(\boldsymbol{\eta}, \mathbf{r}, w, \mu_1, \mu_2) \in \mathbf{V}_h \times \mathbf{V}_h \times W_h \times M_h^\circ \times M_h^\Gamma$. In particular, $\widehat{\varepsilon}_h^u = 0$ on Γ. Taking $\boldsymbol{\eta} = \boldsymbol{\varepsilon}_h^\sigma$, $\mathbf{r} = \boldsymbol{\varepsilon}_h^q$, $w = \varepsilon_h^u$, and $\mu_1 = \widehat{\varepsilon}_h^u$ in the error equations and adding the equations, we reach the energy identity

$$\|\boldsymbol{\varepsilon}_h^\sigma\|_\kappa^2 + |\mathrm{P}\varepsilon_h^u - \widehat{\varepsilon}_h^u|_\tau^2 = (\kappa\mathbf{e}_h^\sigma, \boldsymbol{\varepsilon}_h^\sigma)_{\mathscr{T}_h} + (\mathbf{e}_h^q, \boldsymbol{\varepsilon}_h^\sigma)_{\mathscr{T}_h}.$$

The Cauchy–Schwarz inequality and some simplifications show that

$$\|\boldsymbol{\varepsilon}_h^\sigma\|_\kappa^2 + |\mathrm{P}\varepsilon_h^u - \widehat{\varepsilon}_h^u|_\tau^2 \leqslant \|\mathbf{e}_h^\sigma + \kappa^{-1}\mathbf{e}_h^q\|_\kappa \|\boldsymbol{\varepsilon}_h^\sigma\|_\kappa, \tag{4.22}$$

where $\|\cdot\|_\kappa = \|\kappa^{1/2}\cdot\|_\Omega$. This gives the first group of estimates.

Proposition 4.7 *For the HDG+ approximation* (4.20) *of* (4.19), *we have the estimates*

$$\|\boldsymbol{\varepsilon}_h^\sigma\|_\kappa + |\mathrm{P}(\varepsilon_h^u - \widehat{\varepsilon}_h^u)|_\tau \leqslant 2\|\kappa^{-1/2}\|_{L^\infty}\|\mathbf{e}_h^q\|_\Omega + 2\|\mathbf{e}_h^\sigma\|_\kappa, \tag{4.23a}$$

$$\|\boldsymbol{\varepsilon}_h^q\|_\Omega \leqslant (1 + \|\kappa^{1/2}\|_{L^\infty}\|\kappa^{-1/2}\|_{L^\infty})\|\mathbf{e}_h^q\|_\Omega + 2\|\kappa^{1/2}\|_{L^\infty}\|\mathbf{e}_h^\sigma\|_\kappa. \tag{4.23b}$$

Therefore, the extended HDG+ method is optimally convergent in the approximations for $\boldsymbol{\sigma}$ *and* $\mathbf{q}$.

Proof From (4.22), we easily obtain

$$\|\boldsymbol{\varepsilon}_h^\sigma\|_\kappa \leqslant \|\mathbf{e}_h^\sigma\|_\kappa + \|\kappa^{-1}\mathbf{e}_h^q\|_\kappa. \tag{4.24}$$

This and (4.22) prove (4.23a). On the other hand, from (4.21a) with $\boldsymbol{\eta} = \boldsymbol{\varepsilon}_h^q$, we obtain

$$\|\boldsymbol{\varepsilon}_h^q\|_\Omega \leqslant \|\mathbf{e}_h^q\|_\Omega + \|\kappa^{1/2}\|_{L^\infty}(\|\boldsymbol{\varepsilon}_h^\sigma\|_\kappa + \|\mathbf{e}_h^\sigma\|_\kappa). \tag{4.25}$$

Combining (4.24) and (4.25), we obtain (4.23b).

Duality estimates. To estimate ε_h^u, we again use the duality argument. Consider the adjoint equations of (4.19):

$$\begin{aligned} \kappa\boldsymbol{\psi} - \boldsymbol{\xi} &= \mathbf{0} && \text{in } \Omega, && (4.26a)\\ \boldsymbol{\psi} - \nabla\theta &= \mathbf{0} && \text{in } \Omega, && (4.26b)\\ -\operatorname{div}\boldsymbol{\xi} &= \varepsilon_h^u && \text{in } \Omega, && (4.26c)\\ \theta &= 0 && \text{on } \Gamma. && (4.26d) \end{aligned}$$

As usual, we assume elliptic regularity for (4.26)

$$\|\theta\|_{2,\Omega} + \|\boldsymbol{\psi}\|_{1,\Omega} + \|\boldsymbol{\xi}\|_{1,\Omega} \leqslant C_{\text{reg}}\|\varepsilon_h^u\|_\Omega. \tag{4.27}$$

Proposition 4.8 *Assuming $k \geqslant 1$ and (4.27) holds, then for the HDG+ approximation (4.20) of (4.19), we have the estimates*

$$\|\varepsilon_h^u\|_\Omega \lesssim h(\|\mathbf{e}_h^\sigma\|_\Omega + \|\mathbf{e}_h^q\|_\Omega). \tag{4.28}$$

Proof We first define the dual projection

$$(\boldsymbol{\Pi}\boldsymbol{\xi}, \Pi\theta) = \prod_{K\in\mathcal{T}_h} \Pi^{\text{HDG+}}(\boldsymbol{\xi}, \theta; -\tau),$$

which we complement with $\mathrm{Q}\boldsymbol{\psi}$ (recall that Q is the L^2 orthogonal projection onto $\mathbf{V}_h$). Defining $\mathbf{e}_h^\psi := \mathrm{Q}\boldsymbol{\psi} - \boldsymbol{\psi}$ and $\mathbf{e}_h^\xi := \boldsymbol{\Pi}\boldsymbol{\xi} - \boldsymbol{\xi}$, it is easy to see that

$$\begin{aligned} (\kappa\mathrm{Q}\boldsymbol{\psi}, \boldsymbol{\eta})_{\mathcal{T}_h} - (\boldsymbol{\Pi}\boldsymbol{\xi}, \boldsymbol{\eta})_{\mathcal{T}_h} &= (\kappa\mathbf{e}_h^\psi, \boldsymbol{\eta})_{\mathcal{T}_h} - (\mathbf{e}_h^\xi, \boldsymbol{\eta})_{\mathcal{T}_h}, && (4.29a)\\ (Q\boldsymbol{\psi}, \mathbf{r})_{\mathcal{T}_h} + (\Pi\theta, \nabla\cdot\mathbf{r})_{\mathcal{T}_h} - \langle \mathrm{P}\theta, \mathbf{r}\cdot\mathbf{n}\rangle_{\partial\mathcal{T}_h} &= 0, && (4.29b)\\ -(\nabla\cdot\boldsymbol{\Pi}\boldsymbol{\xi}, w)_{\mathcal{T}_h} + \langle\tau\mathrm{P}(\Pi\theta - \theta), w\rangle_{\partial\mathcal{T}_h} &= (\varepsilon_h^u, w)_{\mathcal{T}_h}, && (4.29c)\\ \langle\boldsymbol{\Pi}\boldsymbol{\xi}\cdot\mathbf{n} - \tau\mathrm{P}(\Pi\theta - \theta), \mu_1\rangle_{\partial\mathcal{T}_h\setminus\Gamma} &= 0, && (4.29d)\\ \langle\mathrm{P}\theta, \mu_2\rangle_\Gamma &= 0, && (4.29e) \end{aligned}$$

for all $(\boldsymbol{\eta}, \mathbf{r}, w, \mu_1, \mu_2) \in \mathbf{V}_h \times \mathbf{V}_h \times W_h \times M_h^\circ \times M_h^\Gamma$,

Testing (4.21) with $\boldsymbol{\eta} = \mathrm{Q}\boldsymbol{\psi}$, $\mathbf{r} = \boldsymbol{\Pi}\boldsymbol{\xi}$, $w = \Pi\theta$, $\mu_1 = \mathrm{P}\theta$, $\mu_2 = -\boldsymbol{\varepsilon}_h^q\cdot\mathbf{n} - \tau(\varepsilon_h^u - \widehat{\varepsilon}_h^u)$, testing (4.29) with $\boldsymbol{\eta} = \boldsymbol{\varepsilon}_h^\sigma$, $\mathbf{r} = \boldsymbol{\varepsilon}_h^q$, $w = \varepsilon_h^u$, $\mu_1 = \widehat{\varepsilon}_h^u$, $\mu_2 = \boldsymbol{\Pi}\boldsymbol{\xi}\cdot\mathbf{n} - \tau\mathrm{P}(\Pi\theta - \theta)$, and comparing the two sets of equations, we obtain

$$(\kappa\mathbf{e}_h^\sigma, \mathrm{Q}\boldsymbol{\psi})_{\mathcal{T}_h} + (\mathbf{e}_h^q, \mathrm{Q}\boldsymbol{\psi})_{\mathcal{T}_h} = (\kappa\mathbf{e}_h^\psi, \boldsymbol{\varepsilon}_h^\sigma)_{\mathcal{T}_h} - (\mathbf{e}_h^\xi, \boldsymbol{\varepsilon}_h^\sigma)_{\mathcal{T}_h} + \|\varepsilon_h^u\|_\Omega^2. \tag{4.30}$$

Notice that

$$\begin{aligned} (\kappa\mathbf{e}_h^\sigma, \mathrm{Q}\boldsymbol{\psi})_{\mathcal{T}_h} - (\kappa\mathbf{e}_h^\psi, \boldsymbol{\varepsilon}_h^\sigma)_{\mathcal{T}_h} &= (\kappa(\boldsymbol{\sigma}_h - \boldsymbol{\sigma}), \mathbf{e}_h^\psi)_{\mathcal{T}_h} + (\kappa\mathbf{e}_h^\sigma, \boldsymbol{\psi})_{\mathcal{T}_h},\\ (\mathbf{e}_h^q, \mathrm{Q}\boldsymbol{\psi})_{\mathcal{T}_h} + (\mathbf{e}_h^\xi, \boldsymbol{\varepsilon}_h^\sigma)_{\mathcal{T}_h} &= (\mathbf{e}_h^q, \mathrm{Q}\boldsymbol{\psi} - \boldsymbol{\psi})_{\mathcal{T}_h} + (\mathbf{e}_h^q, \boldsymbol{\psi})_{\mathcal{T}_h} + (\mathbf{e}_h^\xi, \boldsymbol{\varepsilon}_h^\sigma)_{\mathcal{T}_h}. \end{aligned}$$

Combining the above estimates with (4.30) and fact that $\kappa\boldsymbol{\psi} - \boldsymbol{\xi} = \mathbf{0}$, we have

$$\begin{aligned}\|\varepsilon_h^u\|_\Omega^2 &= (\kappa(\boldsymbol{\sigma}_h - \boldsymbol{\sigma}), \mathrm{Q}\boldsymbol{\psi} - \boldsymbol{\psi})_{\mathcal{T}_h} + (\mathbf{e}_h^\sigma, \boldsymbol{\xi} - \Pi_k\boldsymbol{\xi})_{\mathcal{T}_h} \\ &+ (\mathbf{e}_h^q, \mathrm{Q}\boldsymbol{\psi} - \boldsymbol{\psi})_{\mathcal{T}_h} + (\mathbf{e}_h^q, \boldsymbol{\psi} - \Pi_{k-1}\boldsymbol{\psi})_{\mathcal{T}_h} + (\boldsymbol{\varepsilon}_h^\sigma, \boldsymbol{\Pi}\boldsymbol{\xi} - \boldsymbol{\xi})_{\mathcal{T}_h}.\end{aligned}$$

Now using the regularity assumption (4.27) and Proposition 4.6 with $m = 1$, we have

$$\begin{aligned}\|\boldsymbol{\psi} - \Pi_{k-1}\boldsymbol{\psi}\|_\Omega + \|\mathrm{Q}\boldsymbol{\psi} - \boldsymbol{\psi}\|_\Omega &\lesssim h\|\boldsymbol{\psi}\|_{1,\Omega} \lesssim h\|\varepsilon_h^u\|_\Omega, \\ \|\boldsymbol{\Pi}\boldsymbol{\xi} - \boldsymbol{\xi}\|_\Omega + \|\boldsymbol{\xi} - \Pi_k\boldsymbol{\xi}\|_\Omega &\lesssim h(\|\boldsymbol{\xi}\|_{1,\Omega} + \|\theta\|_{2,\Omega}) \lesssim h\|\varepsilon_h^u\|_\Omega.\end{aligned}$$

The rest of the proof follows easily from Proposition 4.7.

Exercises

1. Prove that (4.20) is uniquely solvable (Hint. Test with $\boldsymbol{\eta} = \boldsymbol{\sigma}_h$, $\mathbf{r} = \mathbf{q}_h$, $w = u_h$, and $\mu = \widehat{u}_h$, and then copy the proof of Proposition 4.1.)

Chapter 5
HDG Methods for Evolutionary Equations

5.1 The Dirichlet Form and the Dirichlet Lifting

In this section we explore a simple form of writing the HDG and HDG+ equations in "primal" form, with u_h as the only unknown. This will shed some light on relations of these methods with similar schemes and will be useful for some arguments on the HDG discretization of evolutionary equations. The presentation of the methods in the coming sections can be done simultaneously for the HDG and HDG+ methods. *We will only give detailed proofs for the HDG+ method and will leave the classical HDG proofs as an exercise.* In the coming sections, we will always have

$$\mathbf{V}_h := \prod_{K\in\mathcal{T}_h} \boldsymbol{\mathcal{P}}_k(K), \qquad M_h := \prod_{e\in\mathcal{E}_h} \mathcal{P}_k(e) = M_h^\circ \oplus M_h^\Gamma,$$

and either (HDG)

$$W_h := \prod_{K\in\mathcal{T}_h} \mathcal{P}_k(K), \qquad \tau \in \mathcal{R}_0(\partial K), \quad \tau \geqslant 0, \quad \tau|_{\partial K} \neq 0 \quad \forall K, \tag{5.1}$$

or (HDG+)

$$W_h := \prod_{K\in\mathcal{T}_h} \mathcal{P}_{k+1}(K), \qquad \tau \in \mathcal{R}_0(\partial K), \quad \tau \equiv h_K^{-1} \quad \text{on } \partial K \quad \forall K. \tag{5.2}$$

We will include the projection P in the definition of the flux

$$\widehat{\mathbf{q}}_h \cdot \mathbf{n} = \mathbf{q}_h \cdot \mathbf{n} + \tau(\mathrm{P}u_h - \widehat{u}_h),$$

S. Du and F.-J. Sayas, *An Invitation to the Theory of the Hybridizable Discontinuous Galerkin Method*, SpringerBriefs in Mathematics,
https://doi.org/10.1007/978-3-030-27230-2_5

even when the choice (5.1) is done. Note that $u_h|_{\partial K} \in \mathscr{R}_k(\partial K)$ and therefore $\mathrm{P}u_h = u_h$ on ∂K.

The HDG gradient. Consider the operator

$$\mathscr{G}_h = (\mathscr{G}_h^q, \mathscr{G}_h^{\widehat{u}}) : W_h \to \mathbf{V}_h \times M_h,$$

given by

$$\mathscr{G}_h u_h = (\mathscr{G}_h^q u_h, \mathscr{G}_h^{\widehat{u}} u_h) = (\mathbf{q}_h, \widehat{u}_h) \in \mathbf{V}_h \times M_h, \tag{5.3a}$$

satisfying

$$(\kappa^{-1}\mathbf{q}_h, \mathbf{r})_{\mathscr{T}_h} - (u_h, \operatorname{div}\mathbf{r})_{\mathscr{T}_h} + \langle \widehat{u}_h, \mathbf{r}\cdot\mathbf{n}\rangle_{\partial\mathscr{T}_h} = 0 \qquad \forall \mathbf{r} \in \mathbf{V}_h, \tag{5.3b}$$

$$-\langle \mathbf{q}_h \cdot \mathbf{n} + \tau(u_h - \widehat{u}_h), \mu\rangle_{\partial\mathscr{T}_h\setminus\Gamma} = 0 \qquad \forall \mu \in M_h^\circ, \tag{5.3c}$$

$$\langle \widehat{u}_h, \mu\rangle_\Gamma = 0 \qquad \forall \mu \in M_h^\Gamma. \tag{5.3d}$$

Note that these are the first, third, and fourth equations of the HDG system (3.1) or (4.3) considering that u_h is known and that the Dirichlet data vanish. Furthermore, we have $\mathscr{G}_h u_h \in \mathbf{V}_h \times M_h^\circ$. This HDG gradient is not the same as the one that can be found in [84]. In fact $\mathscr{G}_h^q u_h \approx -\kappa\nabla u_h$, and therefore the vector component of $\mathscr{G}_h^q u_h$ is just the associated HDG flux inside the element. More on this will be discussed at the end of the section.

Proposition 5.1 *Equations* (5.3) *are uniquely solvable.*

Proof We just need to prove uniqueness of solution of the homogeneous system. Let then $(\mathbf{q}_h, \widehat{u}_h) \in \mathbf{V}_h \times M_h^\circ$ satisfy

$$(\kappa^{-1}\mathbf{q}_h, \mathbf{r})_{\mathscr{T}_h} + \langle \widehat{u}_h, \mathbf{r}\cdot\mathbf{n}\rangle_{\partial\mathscr{T}_h} = 0 \qquad \forall \mathbf{r} \in \mathbf{V}_h,$$

$$-\langle \mathbf{q}_h \cdot \mathbf{n}, \mu\rangle_{\partial\mathscr{T}_h\setminus\Gamma} + \langle \tau\widehat{u}_h, \mu\rangle_{\partial\mathscr{T}_h\setminus\Gamma} = 0 \qquad \forall \mu \in M_h^\circ.$$

Testing with $\mathbf{r} = \mathbf{q}_h$ and $\mu = \widehat{u}_h$, and adding the resulting expressions, we obtain

$$(\kappa^{-1}\mathbf{q}_h, \mathbf{q}_h)_{\mathscr{T}_h} + \langle \tau\widehat{u}_h, \widehat{u}_h\rangle_{\partial\mathscr{T}_h} = 0.$$

This proves that $\mathbf{q}_h = \mathbf{0}$ and $\widehat{u}_h = 0$.

The HDG Dirichlet form. We consider the bilinear form $D_h : W_h \times W_h \to \mathbb{R}$ given by

$$D_h(u_h, w_h) := (\operatorname{div}\mathscr{G}_h^q u_h, w_h)_{\mathscr{T}_h} + \langle \tau(\mathrm{P}u_h - \mathscr{G}_h^{\widehat{u}} u_h), w_h\rangle_{\partial T_h}.$$

Proposition 5.2 *Let* $(\mathbf{q}_h, \widehat{u}_h) = \mathscr{G}_h u_h$ *and* $(\mathbf{p}_h, \widehat{w}_h) = \mathscr{G}_h w_h$ *for* $u_h, w_h \in W_h$. *We can write the Dirichlet form as*

$$D_h(u_h, w_h) = (\kappa^{-1}\mathbf{q}_h, \mathbf{p}_h)_{\mathscr{T}_h} + \langle \tau(\mathrm{P}u_h - \widehat{u}_h), \mathrm{P}w_h - \widehat{w}_h\rangle_{\partial T_h}, \tag{5.4}$$

and therefore D_h is symmetric and positive definite.

Proof In order to show (5.4), we use the first equation defining $\mathscr{G}_h w_h$ and the second equation defining $\mathscr{G}_h u_h$ (see (5.3)), namely,

$$\begin{aligned} (\kappa^{-1}\mathbf{p}_h, \mathbf{r})_{\mathscr{T}_h} + \langle \widehat{w}_h, \mathbf{r}\cdot\mathbf{n}\rangle_{\partial\mathscr{T}_h} &= (w_h, \operatorname{div}\mathbf{r})_{\mathscr{T}_h} && \forall \mathbf{r}\in\mathbf{V}_h,\\ -\langle \mathbf{q}_h\cdot\mathbf{n}, \mu\rangle_{\partial\mathscr{T}_h} + \langle \tau\widehat{u}_h, \mu\rangle_{\partial\mathscr{T}_h} &= \langle \tau \mathrm{P} u_h, \mu\rangle_{\partial\mathscr{T}_h} && \forall \mu\in M_h^\circ. \end{aligned}$$

We test the above with $\mathbf{r} = \mathbf{q}_h$ and $\mu = \widehat{w}_h \in M_h^\circ$ and add the result to obtain (after rearranging terms)

$$(\kappa^{-1}\mathbf{q}_h, \mathbf{p}_h)_{\mathscr{T}_h} - \langle \tau(\mathrm{P}u_h - \widehat{u}_h), \widehat{w}_h\rangle_{\partial\mathscr{T}_h} = (w_h, \operatorname{div}\mathbf{q}_h)_{\mathscr{T}_h}.$$

Since by definition

$$D_h(u_u, w_h) = (\operatorname{div}\mathbf{q}_h, w_h)_{\mathscr{T}_h} + \langle \tau(\mathrm{P}u_u - \widehat{u}_h), \mathrm{P}w_h\rangle_{\partial\mathscr{T}_h},$$

Equation (5.4) follows readily. Symmetry and positive semidefiniteness of D_h follow from (5.4). Finally, if $D_h(u_h, u_h) = 0$, then $\mathbf{q}_h = \mathbf{0}$ and $\mathrm{P}u_h - \widehat{u}_h = 0$ on ∂K for all K (we are dealing with the HDG+ case where $\tau > 0$). The rest of the proof proceeds like the end of the proof of Proposition 4.1 (unique solvability of the HDG+ equations). We have for all $\mathbf{r}\in\mathbf{V}_h$

$$\begin{aligned} (\nabla u_h, \mathbf{r})_{\mathscr{T}_h} &= -(\operatorname{div}\mathbf{r}, u_h)_{\mathscr{T}_h} + \langle u_h, \mathbf{r}\cdot\mathbf{n}\rangle_{\partial\mathscr{T}_h} \\ &= \langle u_h - \widehat{u}_h, \mathbf{r}\cdot\mathbf{n}\rangle_{\partial\mathscr{T}_h} && (\text{by}(5.3\text{b}) \text{ and } \mathbf{q}_h = \mathbf{0})\\ &= \langle \mathrm{P}u_u - \widehat{u}_h, \mathbf{r}\cdot\mathbf{n}\rangle_{\partial\mathscr{T}_h} && (\mathbf{r}\cdot\mathbf{n}\in\mathscr{R}_k(\partial K)\ \ \forall K)\\ &= 0. && (\mathrm{P}u_u - \widehat{u}_h = 0) \end{aligned}$$

Taking $\mathbf{r} = \nabla u_h$, we prove that u_h is constant on each element. Using the fact that $\mathrm{P}u_h = \widehat{u}_h$, that $\widehat{u}_h$ is single-valued on internal faces and vanishes on Γ, it follows that $u_h = 0$. This shows positive definiteness of D_h.

The Dirichlet lifting. Consider the operator $\mathscr{L}_h = (\mathscr{L}_h^q, \mathscr{L}_h^{\widehat{u}}) : L^2(\Gamma) \to \mathbf{V}_h \times M_h$, given by

$$\mathscr{L}_h g = (\mathscr{L}_h^q g, \mathscr{L}_h^{\widehat{u}} g) = (\mathbf{q}_h, \widehat{u}_h) \in \mathbf{V}_h \times W_h, \tag{5.5a}$$

satisfying

$$\begin{aligned} (\kappa^{-1}\mathbf{q}_h, \mathbf{r})_{\mathscr{T}_h} + \langle \widehat{u}_h, \mathbf{r}\cdot\mathbf{n}\rangle_{\partial\mathscr{T}_h} &= 0 && \forall \mathbf{r}\in\mathbf{V}_h, && (5.5\text{b})\\ -\langle \mathbf{q}_h\cdot\mathbf{n}, \mu\rangle_{\partial\mathscr{T}_h\setminus\Gamma} + \langle \tau\widehat{u}_h, \mu\rangle_{\partial\mathscr{T}_h\setminus\Gamma} &= 0 && \forall \mu\in M_h^\circ, && (5.5\text{c})\\ \langle \widehat{u}_h, \mu\rangle_\Gamma &= \langle g, \mu\rangle_\Gamma && \forall \mu\in M_h^\Gamma. && (5.5\text{d}) \end{aligned}$$

Once again, these are the first, third, and fourth equations of the HDG scheme assuming that $u_h = 0$ and solving for the other two variables. These equations are uniquely solvable since the associated system has the same matrix as Eqs. (5.3) defining the

HDG gradient, with a different right-hand side. We finally define the bilinear form $\ell_h : L^2(\Gamma) \times W_h \to \mathbb{R}$, given by

$$\ell_h(g, w_h) := (\operatorname{div} \mathscr{L}_h^q g, w_h)_{\mathscr{T}_h} - \langle \tau \mathscr{L}_h^{\widehat{u}} g, w_h \rangle_{\partial \mathscr{T}_h}.$$

A primal form for the HDG and HDG+ schemes. Using the Dirichlet form and the lifting of Dirichlet conditions, we can rewrite the HDG and HDG+ equations in an equivalent form that involves u_h only. Note that this is not an interesting expression from the computational standpoint, but we will use it for some forthcoming arguments about evolutionary equations. The proof of the following result is left as an exercise to the reader.

Proposition 5.3 *With the respective definitions of D_h and ℓ_h, Eqs.* (3.1) *and* (4.3) *are equivalent to finding $u_h \in W_h$ such that*

$$D_h(u_h, w) = (f, w)_{\mathscr{T}_h} - \ell_h(g, w) \qquad \forall w \in W_h, \tag{5.6a}$$

with the reconstruction

$$(\mathbf{q}_h, \widehat{u}_h) = \mathscr{G}_h u_h + \mathscr{L}_h g. \tag{5.6b}$$

Exercises

1. Use Lemmas 2.1 and 2.2 to show that the HDG gradient is well defined in the classical HDG case.
2. Prove that the Dirichlet form for the classical HDG method is symmetric and positive definite. (Hint. Use ideas from the proof of Proposition 3.1.)
3. Prove Proposition 5.3. (Hint. Show that the unique solution of (5.6a) and the reconstruction (5.6b) solve the HDG equations. This actually proves the equivalence of the systems. Why?)

5.2 Semidiscretization of the Heat Equation

In what follows we will present and analyze HDG (and HDG+) semidiscretizations of the heat and wave equations. As a general rule, we will consider functions of the time variable with values in Sobolev or Lebesgue spaces. Only the time variable will be displayed. Time differentiation will be assumed to be done in a classical way and denoted with upper dots.

The heat equation. Given data $f : [0, \infty) \to L^2(\Omega)$ and $g : [0, \infty) \to H^{1/2}(\Gamma)$, and an initial value $u_0 \in L^2(\Omega)$, we consider the model problem for the heat equation

$$\begin{aligned}
\dot{u}(t) &= \operatorname{div}(\kappa \nabla u(t)) + f(t) && \text{in } \Omega \quad \forall t > 0,\\
u(t) &= g(t) && \text{on } \Gamma \quad \forall t > 0,\\
u(0) &= u_0.
\end{aligned}$$

Introducing the heat flux $\mathbf{q}(t) := -\kappa \nabla u(t)$, we can write the above as a first-order system

$$\kappa^{-1}\mathbf{q}(t) + \nabla u(t) = \mathbf{0} \qquad \text{in } \Omega \quad \forall t > 0, \tag{5.7a}$$
$$\operatorname{div} \mathbf{q}(t) + \dot{u}(t) = f(t) \qquad \text{in } \Omega \quad \forall t > 0, \tag{5.7b}$$
$$u(t) = g(t) \qquad \text{on } \Gamma \quad \forall t > 0, \tag{5.7c}$$
$$u(0) = u_0. \tag{5.7d}$$

HDG semidiscretization. The HDG semidiscretization in space of (5.7) looks for

$$\mathbf{q}_h, u_h, \widehat{u}_h : [0, \infty) \to \mathbf{V}_h \times W_h \times M_h, \tag{5.8a}$$

satisfying for all $t > 0$

$$(\kappa^{-1}\mathbf{q}_h(t), \mathbf{r})_{\mathscr{T}_h} - (u_h(t), \operatorname{div} \mathbf{r})_{\mathscr{T}_h} + \langle \widehat{u}_h(t), \mathbf{r}\cdot\mathbf{n}\rangle_{\partial\mathscr{T}_h} = 0, \tag{5.8b}$$
$$(\operatorname{div} \mathbf{q}_h(t), w)_{\mathscr{T}_h} + (\dot{u}_h(t), w)_{\mathscr{T}_h} + \langle \tau(\mathrm{P}u_h(t) - \widehat{u}_h(t)), w\rangle_{\partial\mathscr{T}_h} = (f(t), w)_{\mathscr{T}_h}, \tag{5.8c}$$
$$-\langle \mathbf{q}_h(t)\cdot\mathbf{n} + \tau(\mathrm{P}u_h(t) - \widehat{u}_h(t)), \mu_1\rangle_{\partial\mathscr{T}_h\setminus\Gamma} = 0, \tag{5.8d}$$
$$\langle \widehat{u}_h(t), \mu_2\rangle_\Gamma = \langle g(t), \mu_2\rangle_\Gamma, \tag{5.8e}$$

for all $(\mathbf{r}, w, \mu_1, \mu_2) \in \mathbf{V}_h \times W_h \times M_h^\circ \times M_h^\Gamma$ and with given initial value $u_h(0)$. We will assume that we have chosen the initial values in the form

$$(\mathbf{q}_h(0), u_h(0)) = \Pi(\mathbf{q}(0), u(0); \tau). \tag{5.8f}$$

Using the HDG Dirichlet form and lifting of Sect. 5.1, we can easily see that (5.8) can be equivalently written as the search for

$$u_h : [0, \infty) \to W_h \tag{5.9a}$$

satisfying for all $t > 0$

$$(\dot{u}_h(t), w)_{\mathscr{T}_h} + D_h(u_h(t), w) = (f(t), w)_{\mathscr{T}_h} - \ell_h(g(t), w) \qquad \forall w \in W_h, \tag{5.9b}$$

with given initial condition $u_h(0)$ and the missing variables reconstructed by the gradient and lifting

$$(\mathbf{q}_h(t), \widehat{u}_h(t)) = \mathscr{G}_h u_h(t) + \mathscr{L}_h g(t). \tag{5.9c}$$

Since (5.9b) is a linear system of first-order differential equations, it is easy to prove that if f and g are continuous in $[0, \infty)$, there exists a unique classical solution to (5.9b). The fields defined in (5.9c) postprocess the Dirichlet data and solution with *steady-state* operators (the operators $\mathscr{G}_h$ and $\mathscr{L}_h$ are independent of the time variable).

Dissipativity. Testing (5.9b) with $w = u_h(t)$, it follows that

$$\frac{1}{2}\frac{\mathrm{d}}{\mathrm{d}t}\|u_h(t)\|_\Omega^2 + D_h(u_h(t), u_h(t)) = (f(t), u_h(t))_{\mathscr{T}_h} - \ell_h(g(t), u_h(t)) \quad \forall t > 0. \tag{5.10}$$

In particular, if $f \equiv 0$ and $g \equiv 0$, using the positive definiteness of the Dirichlet form (Proposition 5.2) it follows that

$$\|u_h(t)\|_\Omega \leqslant \|u_h(0)\|_\Omega \qquad \forall t > 0.$$

The identity can be obtained directly from Eqs. (5.8) as follows: testing the equations with $\mathbf{r} = \mathbf{q}_h(t)$, $w = u_h(t)$, $\mu_1 = \widehat{u}_h(t)$ (restricted to $\partial\mathscr{T}_h \setminus \Gamma$) and $\mu_2 = -\widehat{\mathbf{q}}_h(t)\cdot\mathbf{n}$ (restricted to Γ), and adding the results, we obtain

$$\begin{aligned}\frac{1}{2}\frac{\mathrm{d}}{\mathrm{d}t}\|u_h(t)\|_\Omega^2 &+ \|\mathbf{q}_h(t)\|_{\kappa^{-1}}^2 + |\mathrm{P}u_h(t) - \widehat{u}_h(t)|_\tau^2 \\ &= (f(t), u_h(t))_{\mathscr{T}_h} - \langle g(t), \mathbf{q}_h(t)\cdot\mathbf{n} + \tau(\mathrm{P}u_h(t) - \widehat{u}_h(t))\rangle_\Gamma,\end{aligned} \tag{5.11}$$

which is a rewriting of (5.10) (see Proposition 5.2).

5.2.1 Energy Estimates

Error equations and energy estimates. The first part of the analysis of the HDG semidiscretization in space for the heat equation is very simple: it is reduced to writing the error equations and exploiting an energy identity using a very simple lemma. Let

$$\begin{aligned}\boldsymbol{\varepsilon}_h^q &:= \boldsymbol{\Pi}\mathbf{q} - \mathbf{q}_h : [0,\infty) \to \mathbf{V}_h,\\ \varepsilon_h^u &:= \Pi u - u_h : [0,\infty) \to W_h,\\ \widehat{\varepsilon}_h^u &:= \mathrm{P}u - \widehat{u}_h : [0,\infty) \to M_h,\end{aligned}$$

and

$$\mathbf{e}_h^q := \boldsymbol{\Pi}\mathbf{q} - \mathbf{q} : [0,\infty) \to \mathbf{H}^1(\Omega), \qquad e_h^u := \Pi u - u : [0,\infty) \to H^1(\Omega).$$

Proceeding as usual, we obtain the error equations: for all $t \geqslant 0$

$$(\kappa^{-1}\boldsymbol{\varepsilon}_h^q(t), \mathbf{r})_{\mathscr{T}_h} - (\varepsilon_h^u(t), \operatorname{div}\mathbf{r})_{\mathscr{T}_h} + \langle\widehat{\varepsilon}_h^u(t), \mathbf{r}\cdot\mathbf{n}\rangle_{\partial\mathscr{T}_h} = (\kappa^{-1}\mathbf{e}_h^q(t), \mathbf{r})_{\mathscr{T}_h}, \tag{5.12a}$$

$$(\operatorname{div}\boldsymbol{\varepsilon}_h^q(t), w)_{\mathscr{T}_h} + \langle\tau(\mathrm{P}\varepsilon_h^u(t) - \widehat{\varepsilon}_h^u(t)), w\rangle_{\partial\mathscr{T}_h} + (\dot{\varepsilon}_h^u(t), w)_{\mathscr{T}_h} = (\dot{e}_h^u(t), w)_{\mathscr{T}_h}, \tag{5.12b}$$

$$-\langle \boldsymbol{\varepsilon}_h^q(t)\cdot\mathbf{n} + \tau(\mathrm{P}\varepsilon_h^u(t) - \widehat{\varepsilon}_h^u(t)), \mu\rangle_{\partial\mathcal{T}_h\setminus\Gamma} = 0, \qquad (5.12c)$$

$$\langle \widehat{\varepsilon}_h^u(t), \mu\rangle_\Gamma = 0, \qquad (5.12d)$$

for arbitrary test functions $\mathbf{r} \in \mathbf{V}_h$, $w \in W_h$, and $\mu \in M_h$. Recall that our choice of initial value guaranteed that $\varepsilon_h^u(0) = 0$. Using the fact that $\widehat{\varepsilon}_h^u(t) = 0$ on Γ for all t, testing the above with $\mathbf{r} = \boldsymbol{\varepsilon}_h^q(t)$, $w = \varepsilon_h^u(t)$, and $\mu = \widehat{\varepsilon}_h^u(t)$, and finally adding the results (this has been our modus operandi in all the estimates so far), we obtain the identity

$$\begin{aligned}\|\boldsymbol{\varepsilon}_h^q(t)\|_{\kappa^{-1}}^2 + |\mathrm{P}\varepsilon_h^u(t) - \widehat{\varepsilon}_h^u(t)|_\tau^2 + \frac{1}{2}\frac{\mathrm{d}}{\mathrm{d}t}\|\varepsilon_h^u(t)\|_\Omega^2 \\ = (\kappa^{-1}\boldsymbol{\varepsilon}_h^q(t), \mathbf{e}_h^q(t))_{\mathcal{T}_h} + (\dot{e}_h^u(t), \varepsilon_h^u(t))_{\mathcal{T}_h}.\end{aligned} \qquad (5.13)$$

Differentiating (5.12a) with respect to t and testing it with $\mathbf{r} = \boldsymbol{\varepsilon}_h^q(t)$, while we test (5.12b) with $w = \dot{\varepsilon}_h^u(t)$ and (5.12c) with $\mu = \dot{\widehat{\varepsilon}}_h^u(t)$, we obtain a second identity

$$\begin{aligned}\frac{1}{2}\frac{\mathrm{d}}{\mathrm{d}t}\left(\|\boldsymbol{\varepsilon}_h^q(t)\|_{\kappa^{-1}}^2 + |\mathrm{P}\varepsilon_h^u(t) - \widehat{\varepsilon}_h^u(t)|_\tau^2\right) + \|\dot{\varepsilon}_h^u(t)\|_\Omega^2 \\ = (\kappa^{-1}\boldsymbol{\varepsilon}_h^q(t), \dot{\mathbf{e}}_h^q(t))_{\mathcal{T}_h} + (\dot{e}_h^u(t), \dot{\varepsilon}_h^u(t))_{\mathcal{T}_h}.\end{aligned} \qquad (5.14)$$

Before we extract an error estimate from these identities, let us write here a very simple technical lemma that is often used in the theory of numerical methods for evolutionary equations.

Lemma 5.1 *If $\phi, \gamma, \beta : [0, \infty) \to \mathbb{R}$ are nonnegative continuous functions such that γ is nondecreasing and*

$$\phi(t)^2 \leqslant \gamma(t) + \int_0^t \beta(s)\phi(s)\mathrm{d}s \qquad \forall t \geqslant 0, \qquad (5.15)$$

then

$$\phi(t)^2 \leqslant 2\gamma(t) + \left(\int_0^t \beta(s)\mathrm{d}s\right)^2 \qquad \forall t \geqslant 0.$$

Proof Let t be fixed and consider $t^\star$ such that $\phi(t^\star) = \max_{0\leqslant s\leqslant t}\phi(s)$. It is then clear that

$$\phi(t^\star)^2 \leqslant \gamma(t^\star) + \phi(t^\star)\int_0^{t^\star}\beta(s)\mathrm{d}s \leqslant \gamma(t^\star) + \frac{1}{2}\phi(t^\star)^2 + \frac{1}{2}\left(\int_0^{t^\star}\beta(s)\mathrm{d}s\right)^2,$$

and therefore

$$\phi(t)^2 \leqslant \phi(t^\star)^2 \leqslant 2\gamma(t^\star) + \left(\int_0^{t^\star}\beta(s)\mathrm{d}s\right)^2 \leqslant 2\gamma(t) + \left(\int_0^t \beta(s)\mathrm{d}s\right)^2,$$

which completes the proof.

Proposition 5.4 *For all $t \geqslant 0$,*

$$\|\varepsilon_h^u(t)\|_\Omega^2 + \int_0^t \left(\|\boldsymbol{\varepsilon}_h^q(s)\|_{\kappa^{-1}}^2 + 2|\mathrm{P}\varepsilon_h^u(s) - \widehat{\varepsilon}_h^u(s)|_\tau^2\right) \mathrm{d}s \leqslant 2\int_0^t \|\mathbf{e}_h^q(s)\|_{\kappa^{-1}}^2 \mathrm{d}s + 4\left(\int_0^t \|e_h^u(s)\|_\Omega \mathrm{d}s\right)^2. \tag{5.16}$$

Proof From (5.13), it follows that for all $t > 0$ (this involves using the Cauchy–Schwarz inequality and performing some easy simplifications),

$$\|\boldsymbol{\varepsilon}_h^q(t)\|_{\kappa^{-1}}^2 + 2|\mathrm{P}\varepsilon_h^u(t) - \widehat{\varepsilon}_h^u(t)|_\tau^2 + \frac{\mathrm{d}}{\mathrm{d}t}\|\varepsilon_h^u(t)\|_\Omega^2 \leqslant \|\mathbf{e}_h^q(t)\|_{\kappa^{-1}}^2 + 2\|\dot{e}_h^u(t)\|_\Omega\|\varepsilon_h^u(t)\|_\Omega.$$

Integrating on $(0, t)$, we obtain a bound like (5.15) with

$$\phi(t)^2 := \|\varepsilon_h^u(t)\|_\Omega^2 + \int_0^t \left(\|\boldsymbol{\varepsilon}_h^q(s)\|_{\kappa^{-1}}^2 + 2|\mathrm{P}\varepsilon_h^u(s) - \widehat{\varepsilon}_h^u(s)|_\tau^2\right) \mathrm{d}s,$$

$$\gamma(t) := \int_0^t \|\mathbf{e}_h^q(s)\|_{\kappa^{-1}}^2 \mathrm{d}s,$$

$$\beta(t) := 2\|\dot{e}_h^u(t)\|_\Omega.$$

A direct application of Lemma 5.1 then proves the result.

Proposition 5.5 *For all $t \geqslant 0$*

$$\|\boldsymbol{\varepsilon}_h^q(t)\|_{\kappa^{-1}}^2 + |\mathrm{P}\varepsilon_h^u(t) - \widehat{\varepsilon}_h^u(t)|_\tau^2 + \int_0^t \|\dot{\varepsilon}_h^u(s)\|_\Omega^2 \mathrm{d}s \leqslant 2\|\mathbf{e}_h^q(0)\|_{\kappa^{-1}}^2 + 2\int_0^t \|\dot{e}_h^u(s)\|_\Omega^2 \mathrm{d}s + 4\left(\int_0^t \|\dot{\mathbf{e}}_h^q(s)\|_\Omega \mathrm{d}s\right)^2. \tag{5.17}$$

Proof From (5.14), it is easy to prove that

$$\frac{\mathrm{d}}{\mathrm{d}t}\left(\|\boldsymbol{\varepsilon}_h^q(t)\|_{\kappa^{-1}}^2 + |\mathrm{P}\varepsilon_h^u(t) - \widehat{\varepsilon}_h^u(t)|_\tau^2\right) + \|\dot{\varepsilon}_h^u(t)\|_\Omega^2 \leqslant 2\|\boldsymbol{\varepsilon}_h^q(t)\|_{\kappa^{-1}}\|\dot{\mathbf{e}}_h^q(t)\|_{\kappa^{-1}} + \|\dot{e}_h^u(t)\|_\Omega^2. \tag{5.18}$$

Note now that (5.13) evaluated at $t = 0$ and using that $\varepsilon_h^u(0) = 0$ implies

$$\|\boldsymbol{\varepsilon}_h^q(0)\|_{\kappa^{-1}}^2 + |\mathrm{P}\varepsilon_h^u(0) - \widehat{\varepsilon}_h^u(0)|_\tau^2 = (\kappa^{-1}\mathbf{e}_h^q(0), \boldsymbol{\varepsilon}_h^q(0)) \leqslant \frac{1}{2}\|\boldsymbol{\varepsilon}_h^q(0)\|_{\kappa^{-1}}^2 + \frac{1}{2}\|\mathbf{e}_h^q(0)\|_{\kappa^{-1}}^2,$$

and therefore

$$\|\boldsymbol{\varepsilon}_h^q(0)\|_{\kappa^{-1}}^2 + |\mathrm{P}\varepsilon_h^u(0) - \widehat{\varepsilon}_h^u(0)|_\tau^2 \leqslant \|\mathbf{e}_h^q(0)\|_{\kappa^{-1}}^2. \tag{5.19}$$

Integrating (5.18) on $(0, t)$ and using (5.19), we can obtain a bound like (5.15) with

$$\phi(t)^2 := \|\boldsymbol{\varepsilon}_h^q(t)\|_{\kappa^{-1}}^2 + |\mathrm{P}\varepsilon_h^u(t) - \widehat{\varepsilon}_h^u(t)|_\tau^2 + \int_0^t \|\dot{\varepsilon}_h^u(s)\|_\Omega^2 \mathrm{d}s,$$

$$\gamma(t) := \|\mathbf{e}_h^q(0)\|_{\kappa^{-1}}^2 + \int_0^t \|\dot{e}_h^u(s)\|_\Omega^2 \mathrm{d}s,$$

$$\beta(t) := 2\|\dot{\mathbf{e}}_h^q(t)\|_{\kappa^{-1}}.$$

A direct application of Lemma 5.1 proves the result.

5.2.2 *Estimates by Duality*

In order to estimate $\|\varepsilon_h^u(T)\|_\Omega$ (for a fixed value of T), we consider the adjoint problem, which is a backward heat equation in first-order form:

$$\begin{aligned} \kappa^{-1}\boldsymbol{\xi}(t) - \nabla\theta(t) &= \mathbf{0} && \text{in } \Omega \quad t \in [0, T], && (5.20a)\\ -\operatorname{div}\boldsymbol{\xi}(t) - \dot{\theta}(t) &= 0 && \text{in } \Omega \quad t \in [0, T], && (5.20b)\\ \theta(t) &= 0 && \text{on } \Gamma \quad t \in [0, T], && (5.20c)\\ \theta(T) &= \mathscr{Q}_h\varepsilon_h^u(T). && && (5.20d) \end{aligned}$$

The final value for this adjoint problem, $\mathscr{Q}_h\varepsilon_h^u(T)$, uses the operator $\mathscr{Q}_h$ of Proposition 5.8. The key ingredient for the error analysis by duality is the following identity.

Proposition 5.6 *Let* $k \geqslant 1$, $(\boldsymbol{\xi}, \theta)$ *be the solution of the dual problem (5.20), and* $(\boldsymbol{\Pi}\boldsymbol{\xi}(t), \Pi\theta(t)) := \Pi^{\mathrm{HDG+}}(\boldsymbol{\xi}(t), \theta(t); -\tau)$, *then*

$$\begin{aligned} (\varepsilon_h^u(T), \mathscr{Q}_h\varepsilon_h^u(T))_{\mathscr{T}_h} &= \int_0^T (\kappa^{-1}(\mathbf{q}_h(t) - \mathbf{q}(t)), \boldsymbol{\Pi}\boldsymbol{\xi}(t) - \boldsymbol{\xi}(t))_{\mathscr{T}_h}\mathrm{d}t\\ &+ \int_0^T (\mathbf{e}_h^q(t), \nabla\theta(t) - \mathrm{P}_0\nabla\theta(t))_{\mathscr{T}_h}\mathrm{d}t\\ &+ \int_0^T (\dot{u}_h(t) - \dot{u}(t), \Pi\theta(t) - \theta(t))_{\mathscr{T}_h}\mathrm{d}t\\ &+ \int_0^T (\dot{e}_h^u(t), \theta(t) - \mathrm{P}_0\theta(t))_{\mathscr{T}_h}\mathrm{d}t, \end{aligned}$$

where P_0 *is the orthogonal projection onto piecewise constant functions.*

Proof By definition of the HDG+ projection (note that we are using $-\tau$ as constitutive parameter) and by the adjoint problem (5.20), we have the equations

$$(\kappa^{-1}\boldsymbol{\Pi}\boldsymbol{\xi}(t),\mathbf{r})_{\mathcal{T}_h}+(\Pi\theta(t),\operatorname{div}\mathbf{r})_{\mathcal{T}_h}-\langle \mathrm{P}\theta(t),\mathbf{r}\cdot\mathbf{n}\rangle_{\partial\mathcal{T}_h}=(\kappa^{-1}(\boldsymbol{\Pi}\boldsymbol{\xi}(t)-\boldsymbol{\xi}(t)),\mathbf{r})_{\mathcal{T}_h}, \tag{5.21a}$$

$$-(\operatorname{div}\boldsymbol{\Pi}\boldsymbol{\xi}(t),w)_{\mathcal{T}_h}+\langle\tau(\mathrm{P}\Pi\theta(t)-\theta(t)),w\rangle_{\partial\mathcal{T}_h}=(\dot\theta(t),w)_{\mathcal{T}_h}, \tag{5.21b}$$

$$\langle\boldsymbol{\Pi}\boldsymbol{\xi}(t)\cdot\mathbf{n}-\tau(\mathrm{P}\Pi\theta(t)-\mathrm{P}\theta(t)),\mu\rangle_{\partial\mathcal{T}_h\setminus\Gamma}=0, \tag{5.21c}$$

$$\langle\mathrm{P}\theta(t),\mu\rangle_\Gamma=0, \tag{5.21d}$$

for all $t>0$ and test functions in the usual spaces: $\mathbf{r}\in\mathbf{V}_h$, $w\in W_h$, and $\mu\in M_h$. Testing (5.21)(a–c) with $\mathbf{r}=\boldsymbol{\varepsilon}_h^q(t)$, $w=\varepsilon_h^u(t)$, $\mu=\widehat{\varepsilon}_h^u(t)$ and comparing with the error Eqs. (5.12)(a–c) tested with $\mathbf{r}=\boldsymbol{\Pi}\boldsymbol{\xi}(t)$, $w=\Pi\theta(t)$, and $\mu=\mathrm{P}\theta(t)$ yields the equality

$$\begin{aligned}(\kappa^{-1}(\boldsymbol{\Pi}\boldsymbol{\xi}(t)-\boldsymbol{\xi}(t)),\boldsymbol{\varepsilon}_h^q(t))_{\mathcal{T}_h}+(\dot\theta(t),\varepsilon_h^u(t))_{\mathcal{T}_h}\\=(\dot u_h(t)-\dot u(t),\Pi\theta(t))_{\mathcal{T}_h}+(\kappa^{-1}\mathbf{e}_h^q(t),\boldsymbol{\Pi}\boldsymbol{\xi}(t))_{\mathcal{T}_h}.\end{aligned}$$

Rearranging terms and simplifying, we obtain

$$\begin{aligned}(\dot\theta(t),\varepsilon_h^u(t))_{\mathcal{T}_h}=&(\kappa^{-1}(\boldsymbol{\Pi}\boldsymbol{\xi}(t)-\boldsymbol{\xi}(t)),\mathbf{q}_h(t)-\mathbf{q}(t))_{\mathcal{T}_h}+(\kappa^{-1}\boldsymbol{\xi}(t),\mathbf{e}_h^q(t))_{\mathcal{T}_h}\\&+(\dot u_h(t)-\dot u(t),\Pi\theta(t))_{\mathcal{T}_h}\\=&(\kappa^{-1}(\boldsymbol{\Pi}\boldsymbol{\xi}(t)-\boldsymbol{\xi}(t)),\mathbf{q}_h(t)-\mathbf{q}(t))_{\mathcal{T}_h}+(\nabla\theta(t),\mathbf{e}_h^q(t))_{\mathcal{T}_h}\\&+(\dot u_h(t)-\dot u(t),\Pi\theta(t)-\theta(t))_{\mathcal{T}_h}+(\dot u_h(t)-\dot u(t),\theta(t))_{\mathcal{T}_h}\end{aligned}$$

(we have used that $\nabla\theta=\kappa^{-1}\boldsymbol{\xi}$). The key idea now is to realize that the function

$$\eta(t):=(\theta(t),\varepsilon_h^u(t))_{\mathcal{T}_h}$$

satisfies

$$\eta(0)=(\theta(0),\varepsilon_h^u(0))_{\mathcal{T}_h}=0,\quad \eta(T)=(\theta(T),\varepsilon_h^u(T))_{\mathcal{T}_h}=(\varepsilon_h^u(T),\mathcal{Q}_h\varepsilon_h^u(T))_{\mathcal{T}_h},$$

and

$$\begin{aligned}\dot\eta(t)=&(\dot\theta(t),\varepsilon_h^u(t))_{\mathcal{T}_h}+(\theta(t),\dot\varepsilon_h^u(t))_{\mathcal{T}_h}\\=&(\kappa^{-1}(\boldsymbol{\Pi}\boldsymbol{\xi}(t)-\boldsymbol{\xi}(t)),\mathbf{q}_h(t)-\mathbf{q}(t))_{\mathcal{T}_h}+(\nabla\theta(t),\mathbf{e}_h^q(t))_{\mathcal{T}_h}\\&+(\dot u_h(t)-\dot u(t),\Pi\theta(t)-\theta(t))_{\mathcal{T}_h}+(\dot e_h^u(t),\theta(t))_{\mathcal{T}_h}\\=&(\kappa^{-1}(\boldsymbol{\Pi}\boldsymbol{\xi}(t)-\boldsymbol{\xi}(t)),\mathbf{q}_h(t)-\mathbf{q}(t))_{\mathcal{T}_h}+(\nabla\theta(t)-\mathrm{P}_0\nabla\theta(t),\mathbf{e}_h^q(t))_{\mathcal{T}_h}\\&+(\dot u_h(t)-\dot u(t),\Pi\theta(t)-\theta(t))_{\mathcal{T}_h}+(\dot e_h^u(t),\theta(t)-\mathrm{P}_0\theta(t))_{\mathcal{T}_h},\end{aligned}$$

since for $k\geqslant 1$, $\mathbf{e}_h^q(t)$ and $\dot e_h^u(t)$ are orthogonal to piecewise constant functions. Since

$$(\varepsilon_h^u(T), \mathcal{Q}_h \varepsilon_h^u(T))_{\mathcal{T}_h} = \int_0^T \dot{\eta}(t)\mathrm{d}t,$$

the result follows readily.

More notation. For simplicity, we will write $|\!|\!| u |\!|\!|_p$ with $p \in \{1, 2, \infty\}$ to denote the natural $L^p(0,T; L^2(\Omega))$ or $L^p(0,T; \mathbf{L}^2(\Omega))$ norm, i.e.,

$$|\!|\!| u |\!|\!|_1 := \int_0^T \|u(t)\|_\Omega \mathrm{d}t, \tag{5.22a}$$

$$|\!|\!| u |\!|\!|_2 := \left(\int_0^T \|u(t)\|_\Omega^2 \mathrm{d}t\right)^{1/2}, \tag{5.22b}$$

$$|\!|\!| u |\!|\!|_\infty := \operatorname*{ess\ sup}_{t\in(0,T)} \|u(t)\|_\Omega. \tag{5.22c}$$

Note that T is not made explicit in this notation, but the dependence on T in all inequalities will be made explicit, that is, the symbol $\lesssim$ will hide constants independent of h and T. In this language, Propositions 5.4 and 5.5 can be rewritten (ignoring the terms on the boundary) as

$$|\!|\!| \varepsilon_h^u |\!|\!|_\infty + |\!|\!| \boldsymbol{\varepsilon}_h^q |\!|\!|_2 \lesssim |\!|\!| e_h^u |\!|\!|_1 + |\!|\!| \mathbf{e}_h^q |\!|\!|_2, \tag{5.23a}$$

$$|\!|\!| \dot{\varepsilon}_h^u |\!|\!|_2 + |\!|\!| \boldsymbol{\varepsilon}_h^q |\!|\!|_\infty \lesssim \|\mathbf{e}_h^q(0)\|_\Omega + |\!|\!| \dot{e}_h^u |\!|\!|_2 + |\!|\!| \dot{\mathbf{e}}_h^q |\!|\!|_1. \tag{5.23b}$$

The regularity hypothesis. We will use again a regularity result for an associated elliptic problem to prove optimal convergence for the scalar field. The **regularity hypothesis** we will assume is the same as in the previous chapters: there exists $C_{\text{reg}} > 0$ such that for all $f \in L^2(\Omega)$, the solution of the elliptic problem

$$-\operatorname{div}(\kappa \nabla v) = f \quad \text{in } \Omega,$$
$$v = 0 \quad \text{on } \Gamma,$$

satisfies

$$\|\kappa \nabla v\|_{1,\Omega} + \|v\|_{2,\Omega} \leqslant C_{\text{reg}} \|f\|_\Omega.$$

Proposition 5.7 *Assuming $k \geqslant 1$ and the regularity hypothesis above, we have the bound*

$$\|\varepsilon_h^u(T)\|_\Omega \lesssim h \left(\left(\log \frac{T}{h_{\min}^2}\right)^{1/2} (|\!|\!| \mathbf{e}_h^q |\!|\!|_\infty + |\!|\!| \boldsymbol{\varepsilon}_h^q |\!|\!|_\infty) + \frac{h}{h_{\min}} (|\!|\!| \dot{e}_h^u |\!|\!|_2 + |\!|\!| \dot{\varepsilon}_h^u |\!|\!|_2) \right).$$

Proof Assuming the regularity hypothesis, we have the bound for the solution to (5.20)

$$\|\boldsymbol{\xi}(t)\|_{1,\Omega} + \|\theta(t)\|_{2,\Omega} \lesssim \|\dot{\theta}(t)\|_\Omega \qquad \forall t < T.$$

Using the estimates of the HDG+ projection (Proposition 4.6), we have

$$h\|\boldsymbol{\Pi}\boldsymbol{\xi}(t)-\boldsymbol{\xi}(t)\|_\Omega+\|\Pi\theta(t)-\theta(t)\|_\Omega \lesssim h^2(|\boldsymbol{\xi}(t)|_{1,\Omega}+|\theta(t)|_{2,\Omega})\lesssim h^2\|\dot{\theta}(t)\|_\Omega, \tag{5.24a}$$

while classical approximation estimates show that

$$\|\nabla\theta(t)-\mathrm{P}_0\nabla\theta(t)\|_\Omega \lesssim h|\theta(t)|_{2,\Omega}\lesssim h\|\dot{\theta}(t)\|_\Omega, \tag{5.24b}$$

$$\|\theta(t)-\mathrm{P}_0\theta(t)\|_\Omega \lesssim h\|\nabla\theta(t)\|_\Omega. \tag{5.24c}$$

We now revert to the identity proved in Proposition 5.6 to estimate

$$\begin{aligned}
\tfrac{1}{2}\|\varepsilon_h^u(T)\|_\Omega^2 \leqslant&(\varepsilon_h^u(T),\mathcal{Q}_h\varepsilon_h^u(T))_\Omega && \text{(by (5.25))}\\
\leqslant&\|\kappa^{-1}(\mathbf{q}-\mathbf{q}_h)\|_\infty\|\boldsymbol{\Pi}\boldsymbol{\xi}-\boldsymbol{\xi}\|_1+\|\mathbf{e}_h^q\|_\infty\|\nabla\theta-\mathrm{P}_0\nabla\theta\|_1\\
&+\|\dot{u}_h-\dot{u}\|_2\|\Pi\theta-\theta\|_2+\|\dot{e}_h^u\|_2\|\theta-\mathrm{P}_0\theta\|_2\\
\lesssim&h\|\dot{\theta}\|_1(\|\mathbf{e}_h^q\|_\infty+\|\boldsymbol{\varepsilon}_h^q\|_\infty)\\
&+(h^2\|\dot{\theta}\|_2+h\|\nabla\theta\|_2)(\|\dot{e}_h^u\|_2+\|\dot{\varepsilon}_h^u\|_2) && \text{(by (5.24))}\\
\lesssim&\left(\log\frac{T}{h_{\min}^2}\right)^{1/2}(\|\mathbf{e}_h^q\|_\infty+\|\boldsymbol{\varepsilon}_h^q\|_\infty)h\|\varepsilon_h^u(T)\|_\Omega\\
&+\left(\frac{h}{h_{\min}}+1\right)(\|\dot{e}_h^u\|_2+\|\dot{\varepsilon}_h^u\|_2)h\|\varepsilon_h^u(T)\|_\Omega. && \text{(by Prop. 5.10)}
\end{aligned}$$

This completes the proof.

We remark that the terms $\|\boldsymbol{\varepsilon}_h^q\|_\infty$ and $\|\dot{\varepsilon}_h^u\|_2$ in Proposition 5.7 have already been bounded in (5.23b).

5.2.3 A Little Technical Trick

Our goal is the proof of the following result.

Proposition 5.8 *There exists an operator* $\mathcal{Q}_h : L^2(\Omega)\to L^2(\Omega)$ *such that*

(a) $\|\mathcal{Q}_h w\|_\Omega \leqslant \|w\|_\Omega$ *for all* $w\in L^2(\Omega)$.
(b) $\mathcal{Q}_h w\in H_0^1(\Omega)$ *for all* $w\in W_h$ *and*

$$h_{\min}\|\nabla\mathcal{Q}_h w\|_\Omega \lesssim \|w\|_\Omega \qquad \forall w\in W_h.$$

(c) $\|w-\mathcal{Q}_h w\|_\Omega \leqslant \frac{1}{2}\|w\|_\Omega$ *for all* $w\in W_h$.

This operator is constructed in [15] by refinement and projection. We will construct here a simpler version by multiplication by a discrete function in a finer grid. Note

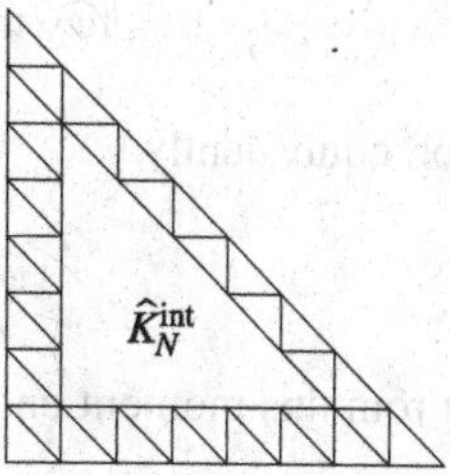

Fig. 5.1 The interior set $\widehat{K}_N^{\rm int}$ of $\widehat{K}$ when $N = 8$

that (c) above implies

$$\|w\|_\Omega^2 = (w, w - \mathcal{Q}_h w)_\Omega + (\mathcal{Q}_h w, w)_\Omega \leqslant \tfrac{1}{2}\|w\|_\Omega^2 + (\mathcal{Q}_h w, w)_\Omega \quad \forall w \in W_h.$$

and therefore

$$\tfrac{1}{2}\|w\|_\Omega^2 \leqslant (w, \mathcal{Q}_h w)_\Omega \qquad \forall w \in W_h. \tag{5.25}$$

A cutoff function in the reference element. Let $\widehat{\mathcal{T}}_N$ be a "uniform" partition of $\widehat{K}$ into smaller simplices with vertices on the principal lattice $\{(i_1/N, \ldots, i_d/N) : i_\ell \geqslant 0 \;\; \forall \ell, \;\; i_1 + \ldots i_d \leqslant N\}$. We consider the interior set (see Fig. 5.1)

$$\widehat{K}_N^{\rm int} := \{\mathbf{x} \in \widehat{K} \, : \, x_i \geqslant 1/N \;\forall i, \;\; 1 - x_1 - \ldots - x_d \geqslant 1/N\},$$

which is also the union of the closures of the elements of $\widehat{\mathcal{T}}_N$ whose boundaries do not intersect with $\partial \hat{K}$. We can then define the function

$$\widehat{\varphi}_N \in H_0^1(\widehat{K}) \bigcap \prod_{T \in \widehat{\mathcal{T}}_N} \mathcal{P}_1(T) \qquad \widehat{\varphi}_N \equiv 1 \;\; \text{in } \widehat{K}_N^{\rm int},$$

or equivalently, $\widehat{\varphi}_N$ is the only continuous piecewise linear function defined on the partition $\widehat{\mathcal{T}}_N$ which takes the unit value on all interior vertices of the mesh and zero on all the boundary vertices. Now consider the functions

$$L^2(\widehat{K}) \ni u \longmapsto \phi_N(u) := \|u\|_{\widehat{K} \setminus \widehat{K}_N^{\rm int}},$$

and note that

$$\|\widehat{\varphi}_N u - u\|_{\widehat{K}} \leqslant \phi_N(u) \qquad \forall u \in L^2(\widehat{K}).$$

It is clear that the sequence of functions $\{\phi_N\}$ is nonincreasing and that $\phi_N(u) \to 0$ for all u. Therefore,

$$\phi_N \to 0 \;\; \text{uniformly in } \{u \in \mathcal{P}_{k+1}(\widehat{K}) \, : \, \|u\|_{\widehat{K}} \leqslant 1\}.$$

This can be proved by Dini's theorem, since the set $\{u \in \mathcal{P}_{k+1}(\widehat{K}) \, : \, \|u\|_{\widehat{K}} \leqslant 1\}$ is compact in $L^2(\widehat{K})$. In any case, this proves that there exists $N = N_k$ such that

$$\|\widehat{\varphi}_N u - u\|_{\widehat{K}} \leqslant \tfrac{1}{2} \qquad \forall u \in \mathscr{P}_{k+1}(\widehat{K}) \quad \|u\|_{\widehat{K}} \leqslant 1,$$

or, equivalently,

$$\|\widehat{\varphi}_N u - u\|_{\widehat{K}} \leqslant \tfrac{1}{2}\|u\|_{\widehat{K}} \qquad \forall u \in \mathscr{P}_{k+1}(\widehat{K}). \tag{5.26}$$

From this moment on, we keep this N fixed for the rest of the argument.

A cut-and-glue process. Consider the function $\varphi_h : \Omega \to \mathbb{R}$ given by

$$\varphi_h|_K = \widehat{\varphi}_N \circ \mathrm{F}_K^{-1} \qquad \forall K \in \mathscr{T}_h.$$

It is clear that $\varphi_h|_K \in H_0^1(K)$ for every K, $0 \leqslant \varphi_h \leqslant 1$ and φ_h is piecewise linear in the refined partition $\mathscr{T}_{h/N}$, obtained by mapping the partition $\mathscr{T}_N$ to each of the elements of $\mathscr{T}_h$. Defining the bounded linear operator in $L^2(\Omega)$

$$\mathscr{Q}_h u := \varphi_h\, u,$$

we have that this operator satisfies (a) in Proposition 5.8. To prove (c) we work element by element, going to the reference element: if $w \in \mathscr{P}_{k+1}(K)$, then

$$\begin{aligned}
\|w - \mathscr{Q}_h w\|_K &= \|w - \varphi_h\, w\|_K \\
&= |J|^{-1/2}\|\widehat{w} - \widehat{\varphi}_N\, \widehat{w}\|_{\widehat{K}} && \text{(by (1.5a) and } \varphi_h \circ \mathrm{F}_K = \widehat{\varphi}_N) \\
&\leqslant \tfrac{1}{2}|J|^{-1/2}\|\widehat{w}\|_{\widehat{K}} && (\widehat{w} \in \mathscr{P}_{k+1}(\widehat{K}) \text{ and (5.26))} \\
&= \tfrac{1}{2}\|w\|_K. && \text{(by (1.5a))}
\end{aligned}$$

Adding over all triangles, it is clear that (c) follows. Finally, note that if $w \in W_h$, then $\varphi_h w$ is continuous and piecewise polynomial of degree $k+2$ on a shape-regular refinement of $\mathscr{T}_h$ and that the refinement depth is fixed by the constant N_k ensuring (5.26). This means that an inverse inequality holds by a simple scaling argument. This proves (b), which completes the proof of Proposition 5.8.

5.2.4 *Regularity Estimates for Parabolic Equations*

Here, we give some direct and simple proofs of known estimates for the solution of parabolic problems and the corresponding corollaries assuming **elliptic regularity**. Let v be the solution of the backward parabolic problem

$$\operatorname{div}(\kappa \nabla v)(t) + \dot{v}(t) = 0 \qquad \text{in } \Omega \quad t \in [0, T], \tag{5.27a}$$
$$v(t) = 0 \qquad \text{on } \Gamma \quad t \in [0, T], \tag{5.27b}$$
$$v(T) = v_T, \tag{5.27c}$$

with $v_T \in H_0^1(\Omega)$.

Proposition 5.9 *If v is the solution to (5.27), then*

$$\|\kappa^{1/2}\nabla v\|_2 \leqslant \frac{1}{\sqrt{2}}\|v_T\|_\Omega, \tag{5.28a}$$

$$\|\mathrm{div}(\kappa\nabla v)\|_2 \leqslant \frac{1}{\sqrt{2}}\|\kappa^{1/2}\nabla v_T\|_\Omega, \tag{5.28b}$$

$$\|(T-\cdot)^{1/2}\mathrm{div}(\kappa\nabla v)\|_2 \leqslant \frac{1}{\sqrt{2\lambda_{\min}}}\|v_T\|_\Omega, \tag{5.28c}$$

where $\lambda_{\min}$ *is the smallest Dirichlet eigenvalue for the diffusion operator* $u \mapsto -\mathrm{div}(\kappa\nabla u)$.

Proof Let $\{(\lambda_n;\phi_n)\}$ be an orthonormal Dirichlet eigensystem for the diffusion operator, that is, $\{\phi_n\}$ is a complete $L^2(\Omega)$ orthonormal sequence and

$$-\,\mathrm{div}(\kappa\nabla\phi_n) = \lambda_n\,\phi_n \qquad \lambda_n \in H_0^1(\Omega) \qquad \forall n. \tag{5.29}$$

A simple computation (this is plain vanilla space–time separation of variables) shows that

$$v(t) = \sum_{n=1}^{\infty} e^{\lambda_n(t-T)} c_n\phi_n \qquad c_n := (\phi_n, v_T)_\Omega.$$

For each value of $t \in [0,T]$, this is an orthogonal series in $L^2(\Omega)$ and its convergence can be shown to be uniform in t using Dini's theorem applied to the functions

$$t \longmapsto \left\| v(t) - \sum_{n=1}^{N} e^{\lambda_n(t-T)} c_n\phi_n \right\|_\Omega^2 = \sum_{n=N+1}^{\infty} e^{2\lambda_n(t-T)}c_n^2.$$

Using the variational formulation of the eigenvalue problem, it follows that

$$(\kappa\nabla v(t), \nabla v(t))_\Omega = \|\kappa^{1/2}\nabla v(t)\|_\Omega^2 = \sum_{n=1}^{\infty}\lambda_n e^{2\lambda_n(t-T)}c_n^2.$$

This series is again uniformly convergent for $t\in[0,T]$ (the fact that $v_T \in H_0^1(\Omega)$ plays a key role here, although this condition can be avoided to prove (5.28a), which can easily be proved by using Dini's theorem again). Therefore

$$\begin{aligned}\int_0^T (\kappa\nabla v(t), \nabla v(t))_\Omega \mathrm{d}t &= \sum_{n=1}^{\infty} c_n^2\lambda_n \int_0^T e^{2\lambda_n(t-T)}\mathrm{d}t\\ &= \sum_{n=1}^{\infty}\frac{c_n^2}{2}(1-e^{-2\lambda_n T}) \leqslant \frac{1}{2}\|v_T\|_\Omega^2,\end{aligned}$$

which proves (5.28a). To prove (5.28b), use (5.29) in the series expansion of v to obtain

$$-\mathrm{div}(\kappa\nabla v)(t) = \sum_{n=1}^{\infty} e^{\lambda_n(t-T)} c_n \lambda_n \phi_n \qquad t < T,$$

and then compute the $L^2(\Omega)$ norm of this orthogonal series

$$\|\mathrm{div}(\kappa\nabla v)(t)\|_\Omega^2 = \sum_{n=1}^{\infty} \lambda_n^2 c_n^2 e^{2\lambda_n(t-T)}.$$

The series in the right-hand side of the above expression converges uniformly in intervals $[0, T-\delta]$ and therefore we can estimate

$$\begin{aligned}\int_0^{T-\delta} \|\mathrm{div}(\kappa\nabla v)(t)\|_\Omega^2 \mathrm{d}t &= \sum_{n=1}^{\infty} c_n^2\lambda_n^2 \int_0^{T-\delta} e^{2\lambda_n(t-T)}\mathrm{d}t \leqslant \sum_{n=1}^{\infty} c_n^2\lambda_n^2 \int_{-\infty}^{T} e^{2\lambda_n(t-T)}\mathrm{d}t \\ &= \frac{1}{2}\sum_{n=1}^{\infty} c_n^2\lambda_n = \frac{1}{2}\|\kappa^{1/2}\nabla v_T\|_\Omega^2,\end{aligned}$$

which proves (5.28b) (after taking the limit $\delta \to 0$). Finally, the same argument shows that

$$\begin{aligned}\int_0^{T-\delta} (T-t)\|\mathrm{div}(\kappa\nabla v)(t)\|_\Omega^2 \mathrm{d}t &= \sum_{n=1}^{\infty} c_n^2\lambda_n^2 \int_0^{T-\delta} (T-t)e^{2\lambda_n(t-T)}\mathrm{d}t \\ &\leqslant \sum_{n=1}^{\infty} c_n^2\lambda_n^2 \int_0^{T} (T-t)e^{2\lambda_n(t-T)}\mathrm{d}t \\ &\leqslant \sum_{n=1}^{\infty} \frac{c_n^2}{2\lambda_n} \leqslant \frac{1}{2\lambda_{\min}}\|v_T\|_\Omega^2,\end{aligned}$$

which completes the proof.

Proposition 5.10 *If v is the solution to (5.27) with $v_T := \mathscr{Q}_h w$ for some $w \in W_h$, then*

$$\|\nabla v\|_2 + h_{\min}\|\dot v\|_2 \lesssim \|w\|_\Omega, \tag{5.30a}$$

$$\|\dot v\|_1 \lesssim \left(\log(T h_{\min}^{-2})\right)^{1/2} \|w\|_\Omega. \tag{5.30b}$$

Proof The estimates (5.30a) are simple consequences of Proposition 5.9: for the first one we bound

$$\|\nabla v\|_2 \lesssim \|\kappa^{1/2}\nabla v\|_2 \lesssim \|\mathscr{Q}_h w\|_\Omega \lesssim \|w\|_\Omega,$$

and for the second one

$$\|\dot{v}\|_2 = \|\mathrm{div}(\kappa\nabla v)\|_2 \lesssim \|\nabla \mathcal{Q}_h w\|_\Omega \lesssim h_{\min}^{-1}\|w\|_\Omega.$$

(We have used the estimates of Proposition 5.8 when dealing with the operator $\mathcal{Q}_h$.) Finally, for arbitrary $\delta > 0$, we estimate

$$\begin{aligned}
&\|\mathrm{div}(\kappa\nabla v)\|_1 \\
&= \int_0^{T-\delta} \|\mathrm{div}(\kappa\nabla v)(t)\|_\Omega \mathrm{d}t + \int_{T-\delta}^{T} \|\mathrm{div}(\kappa\nabla v)(t)\|_\Omega \mathrm{d}t \\
&\leqslant \left(\int_0^{T-\delta} (T-t)^{-1}\mathrm{d}t\right)^{1/2} \left(\int_0^{T-\delta} (T-t)\|\mathrm{div}(\kappa\nabla v)(t)\|_\Omega^2 \mathrm{d}t\right)^{1/2} \\
&\quad + \left(\int_{T-\delta}^{T} \mathrm{d}t\right)^{1/2} \left(\int_{T-\delta}^{T} \|\mathrm{div}(\kappa\nabla v)(t)\|_\Omega^2 \mathrm{d}t\right)^{1/2} && \text{(Cauchy–Schwarz)} \\
&\leqslant \left(\log\frac{T}{\delta}\right)^{1/2} \|(T-\cdot)^{1/2}\mathrm{div}(\kappa\nabla v)\|_2 + \delta^{1/2}\|\mathrm{div}(\kappa\nabla v)\|_2 \\
&\leqslant \left(\log\frac{T}{\delta}\right)^{1/2} \frac{1}{\sqrt{2\lambda_{\min}}}\|\mathcal{Q}_h w\|_\Omega + \frac{\delta^{1/2}}{\sqrt{2}}\|\kappa^{1/2}\nabla\mathcal{Q}_h w\|_\Omega && \text{(Prop. 5.9)} \\
&\lesssim \left(\left(\log\frac{T}{\delta}\right)^{1/2} + \delta^{1/2}h_{\min}^{-1}\right)\|w\|_\Omega. && \text{(Prop. 5.8)}
\end{aligned}$$

Note that the hidden constant is related to an upper bound for κ, to $\lambda_{\min}$ (the lowest Dirichlet eigenvalue of the diffusion operator) and to the constant in Proposition 5.8(b). The proof of (5.30b) is finished by taking $\delta = h_{\min}^2$.

Exercises

1. Let V be any finite-dimensional subspace of $L^2(\widehat{K})$ and let $\{\widehat{\varphi}_N\}$ be the sequence of functions defined in Sect. 5.2.3. Using a basis for V, give a direct proof of the fact that there exists $N = N_V$ such that

$$\|\widehat{\varphi}_N\, u - u\|_{\widehat{K}} \leqslant \tfrac{1}{2}\|u\|_{\widehat{K}} \qquad \forall u \in V.$$

5.3 Semidiscretization of the Wave Equation

The wave equation. Given data $f : [0,\infty) \to L^2(\Omega)$ and $g : [0,\infty) \to H^{1/2}(\Gamma)$, and initial values $u_0, v_0 : \Omega \to \mathbb{R}$, we consider the model problem for the wave equation

$$\begin{aligned}
\rho\, \ddot{u}(t) &= \operatorname{div}(\kappa \nabla u(t)) + f(t) && \text{in } \Omega \quad \forall t > 0,\\
u(t) &= g(t) && \text{on } \Gamma \quad \forall t > 0,\\
u(0) &= u_0,\\
\dot{u}(0) &= v_0.
\end{aligned}$$

The coefficients $\kappa, \rho \in L^\infty(\Omega)$ are assumed to be strictly positive. For convenience, we will often write $\|u\|_\rho := (\rho u, u)_\Omega^{1/2}$. Introducing the heat flux $\mathbf{q}(t) := -\kappa \nabla u(t)$, we can write the above as a first-order system

$$\kappa^{-1}\mathbf{q}(t) + \nabla u(t) = \mathbf{0} \quad \text{in } \Omega \quad \forall t > 0, \tag{5.31a}$$

$$\operatorname{div} \mathbf{q}(t) + \rho \ddot{u}(t) = f(t) \quad \text{in } \Omega \quad \forall t > 0, \tag{5.31b}$$

$$u(t) = g(t) \quad \text{on } \Gamma \quad \forall t > 0, \tag{5.31c}$$

$$u(0) = u_0, \tag{5.31d}$$

$$\dot{u}(0) = v_0. \tag{5.31e}$$

This means that we will be using a *second order in time, first order in space* formulation. There is an alternative option of using a *first order in space and time* formulation. This has been explored in [54]. The current formulation has the advantage of providing an energy conservative scheme.

HDG semidiscretization. The HDG semidiscretization in space of (5.31) looks for

$$\mathbf{q}_h, u_h, \widehat{u}_h : [0,\infty) \to \mathbf{V}_h \times W_h \times M_h, \tag{5.32a}$$

satisfying for all $t > 0$

$$(\kappa^{-1}\mathbf{q}_h(t), \mathbf{r})_{\mathscr{T}_h} - (u_h(t), \operatorname{div} \mathbf{r})_{\mathscr{T}_h} + \langle \widehat{u}_h(t), \mathbf{r}\cdot\mathbf{n}\rangle_{\partial\mathscr{T}_h} = 0, \tag{5.32b}$$

$$\begin{aligned}(\operatorname{div} \mathbf{q}_h(t), w)_{\mathscr{T}_h} &+ (\rho \ddot{u}_h(t), w)_{\mathscr{T}_h}\\ &+ \langle \tau(\mathrm{P}u_h(t) - \widehat{u}_h(t)), w\rangle_{\partial\mathscr{T}_h} = (f(t), w)_{\mathscr{T}_h},\end{aligned} \tag{5.32c}$$

$$-\langle \mathbf{q}_h(t)\cdot\mathbf{n} + \tau(\mathrm{P}u_h(t) - \widehat{u}_h(t)), \mu_1\rangle_{\partial\mathscr{T}_h\setminus\Gamma} = 0, \tag{5.32d}$$

$$\langle \widehat{u}_h(t), \mu_2\rangle_\Gamma = \langle g(t), \mu_2\rangle_\Gamma, \tag{5.32e}$$

for all $(\mathbf{r}, w, \mu_1, \mu_2) \in \mathbf{V}_h \times W_h \times M_h^\circ \times M_h^\Gamma$ and with given initial values $u_h(0)$, $\dot{u}_h(0)$. Fixing initial value approximations that provide optimal error estimates requires a little work. Note that when $t = 0$, Eqs. (5.31) imply that

$$\kappa^{-1}\mathbf{q}(0) + \nabla u(0) = \mathbf{0} \quad \text{in } \Omega, \tag{5.33a}$$

$$\operatorname{div} \mathbf{q}(0) = -\operatorname{div}(\kappa \nabla u_0) =: \lambda \quad \text{in } \Omega, \tag{5.33b}$$

$$u(0) = g(0) \quad \text{on } \Gamma. \tag{5.33c}$$

We now use the HDG+ discretization of (5.33), to find $(\mathbf{q}_h(0), u_h(0), \widehat{u}_h(0)) \in \mathbf{V}_h \times W_h \times M_h$ such that

$$(\kappa^{-1}\mathbf{q}_h(0), \mathbf{r})_{\mathscr{T}_h} - (u_h(0), \operatorname{div}\mathbf{r})_{\mathscr{T}_h} + \langle \widehat{u}_h(0), \mathbf{r}\cdot\mathbf{n}\rangle_{\partial\mathscr{T}_h} = 0, \tag{5.34a}$$

$$(\operatorname{div}\mathbf{q}_h(0), w)_{\mathscr{T}_h} + \langle \tau \mathrm{P}(u_h(0) - \widehat{u}_h(0)), w\rangle_{\partial\mathscr{T}_h} = (\lambda, w)_{\mathscr{T}_h}, \tag{5.34b}$$

$$-\langle \mathbf{q}_h(0)\cdot\mathbf{n} + \tau\mathrm{P}(u_h(0) - \widehat{u}_h(0)), \mu_1\rangle_{\partial\mathscr{T}_h\setminus\Gamma} = 0, \tag{5.34c}$$

$$\langle \widehat{u}_h(0), \mu_2\rangle_\Gamma = \langle g(0), \mu_2\rangle_\Gamma, \tag{5.34d}$$

for all $(\mathbf{r}, w, \mu_1, \mu_2) \in \mathbf{V}_h \times W_h \times M_h^\circ \times M_h^\Gamma$. The component $u_h(0)$ of the solution of (5.34) is used as the first initial condition for the semidiscrete system. We then fix $\dot{u}_h(0)$ using the HDG+ projection in the following form:

$$(\times, \dot{u}_h(0)) = \Pi(-\kappa\nabla v_0, v_0; \tau).$$

Once again, we can use the discrete Dirichlet form and lifting of Sect. 5.1, to show that (5.31) is equivalent to a second-order system of ordinary differential equations

$$u_h : [0, \infty) \to W_h \tag{5.35a}$$

satisfying for all $t > 0$

$$(\rho\ddot{u}_h(t), w)_{\mathscr{T}_h} + D_h(u_h(t), w) = (f(t), w)_{\mathscr{T}_h} - \ell_h(g(t), w) \qquad \forall w \in W_h, \tag{5.35b}$$

with given initial conditions $u_h(0)$, $\dot{u}_h(0)$. The remaining variables are reconstructed with the formula

$$(\mathbf{q}_h(t), \widehat{u}_h(t)) = \mathscr{G}_h u_h(t) + \mathscr{L}_h g(t). \tag{5.35c}$$

The above primal formulation shows that (5.32) has a unique classical solution for continuous-in-time data functions f and g.

Conservation of energy. Testing (5.35b) with $w = \dot{u}_h(t)$, we easily show that

$$\frac{\mathrm{d}}{\mathrm{d}t}\left(\frac{1}{2}\|\dot{u}_h(t)\|_\rho^2 + \frac{1}{2}D_h(u_h(t), u_h(t))\right) = (f(t), \dot{u}_h(t))_{\mathscr{T}_h} - \ell_h(g(t), \dot{u}_h(t)).$$

Therefore, if $f \equiv 0$ and $g \equiv 0$, the energy

$$\frac{1}{2}\|\dot{u}_h(t)\|_\rho^2 + \frac{1}{2}D_h(u_h(t), u_h(t))$$

is constant over time. In more explicit terms (using Proposition 5.2 to get rid of the Dirichlet form), we can prove that with zero right-hand sides (sources and boundary conditions), the following energy function

$$\frac{1}{2}\|\dot{u}_h(t)\|_\rho^2 + \frac{1}{2}\|\mathbf{q}_h(t)\|_{\kappa^{-1}}^2 + \frac{1}{2}|\mathrm{P}u_h(t) - \widehat{u}_h(t)|_\tau^2 \tag{5.36}$$

is constant in time.

5.3.1 Energy Estimates

Error equations. We define $(\boldsymbol{\Pi}\mathbf{q}(t), \Pi u(t)) = \Pi^{\mathrm{HDG+}}(\mathbf{q}(t), u(t); \tau)$, and

$$\boldsymbol{\varepsilon}_h^q(t) = \boldsymbol{\Pi}\mathbf{q}(t) - \mathbf{q}_h(t), \qquad \varepsilon_h^u(t) = \Pi u(t) - u_h(t), \qquad \widehat{\varepsilon}_h^u(t) = \mathrm{P}u(t) - \widehat{u}_h(t), \tag{5.37}$$

$$\mathbf{e}_h^q(t) = \boldsymbol{\Pi}\mathbf{q}(t) - \mathbf{q}(t), \qquad e_h^u(t) = \Pi u(t) - u(t), \tag{5.38}$$

as measurements of the evolution of the semidiscretization error and the approximation error as the solution of (5.31) evolves in time. Not surprisingly, given the fact that we have used the same discretization in space technique, the error equations for the HDG semidiscretization of the wave equation can be proved to be the following slight modification of (5.12): for all $t \geqslant 0$

$$(\kappa^{-1}\boldsymbol{\varepsilon}_h^q(t), \mathbf{r})_{\mathscr{T}_h} - (\varepsilon_h^u(t), \operatorname{div}\mathbf{r})_{\mathscr{T}_h} + \langle \widehat{\varepsilon}_h^u(t), \mathbf{r}\cdot\mathbf{n}\rangle_{\partial\mathscr{T}_h} = (\kappa^{-1}\mathbf{e}_h^q(t), \mathbf{r})_{\mathscr{T}_h}, \tag{5.39a}$$

$$(\rho\ddot{\varepsilon}_h^u(t), w)_{\mathscr{T}_h} + (\operatorname{div}\boldsymbol{\varepsilon}_h^q(t), w)_{\mathscr{T}_h} + \langle \tau(\mathrm{P}\varepsilon_h^u(t) - \widehat{\varepsilon}_h^u(t)), w\rangle_{\partial\mathscr{T}_h} = (\rho\ddot{e}_h^u(t), w)_{\mathscr{T}_h}, \tag{5.39b}$$

$$-\langle \boldsymbol{\varepsilon}_h^q(t)\cdot\mathbf{n} + \tau(\mathrm{P}\varepsilon_h^u(t) - \widehat{\varepsilon}_h^u(t)), \mu\rangle_{\partial\mathscr{T}_h\setminus\Gamma} = 0, \tag{5.39c}$$

$$\langle \widehat{\varepsilon}_h^u(t), \mu\rangle_\Gamma = 0, \tag{5.39d}$$

for arbitrary test functions $\mathbf{r} \in \mathbf{V}_h$, $w \in W_h$, and $\mu \in M_h$. By choice of the discrete initial conditions, we also have

$$\varepsilon_h^u(0) = \Pi u_0 - u_h(0), \qquad \dot{\varepsilon}_h^u(0) = 0. \tag{5.39e}$$

Proposition 5.11 (An estimate at the initial time) *The error at the initial time can be estimated by*

$$\begin{aligned}
\|\boldsymbol{\varepsilon}_h^q(0)\|_{\kappa^{-1}}^2 + |\mathrm{P}\varepsilon_h^u(0) - \widehat{\varepsilon}_h^u(0)|_\tau^2 &\leqslant \|\mathbf{e}_h^q(0)\|_{\kappa^{-1}}^2,\\
\|\dot{\boldsymbol{\varepsilon}}_h^q(0)\|_{\kappa^{-1}}^2 + |\mathrm{P}\dot{\varepsilon}_h^u(0) - \widehat{\dot{\varepsilon}}_h^u(0)|_\tau^2 &\leqslant \|\dot{\mathbf{e}}_h^q(0)\|_{\kappa^{-1}}^2,\\
\|\ddot{\varepsilon}_h^u(0)\|_\rho &\leqslant \|\ddot{e}_h^u(0)\|_\rho.
\end{aligned}$$

Proof The first estimate is the standard error estimate of the elliptic problem (5.33) by its HDG+ approximation (5.34) (see (4.15)). For the second inequality, we differentiate the error Eqs. (5.39) (except the initial conditions) with respect to time, evaluate them at $t = 0$, and test with $\mathbf{r} = \dot{\boldsymbol{\varepsilon}}_h^q(0)$, $w = \dot{\varepsilon}_h^u(0) = 0$, $\mu = \dot{\varepsilon}_h^{\widehat{u}}(0)$. We thus prove

$$\|\dot{\boldsymbol{\varepsilon}}_h^q(0)\|_{\kappa^{-1}}^2 + |\mathrm{P}\dot{\varepsilon}_h^u(0) - \widehat{\dot{\varepsilon}}_h^u(0)|_\tau^2 = (\kappa^{-1}\dot{\mathbf{e}}_h^q(0), \dot{\boldsymbol{\varepsilon}}_h^q(0))_{\mathscr{T}_h},$$

which implies the result. For the final inequality, we use that

$$(\operatorname{div} \boldsymbol{\varepsilon}_h^q(0), w)_{\mathscr{T}_h} + \langle \tau(\mathrm{P}\varepsilon_h^u(0) - \widehat{\varepsilon}_h^u(0)), w\rangle_{\partial\mathscr{T}_h} = 0$$

(this follows from (5.33) and (5.34)), and substitute in (5.39b) at $t = 0$ to prove that

$$(\rho\ddot{\varepsilon}_h^u(0), w)_{\mathscr{T}_h} = (\rho\ddot{e}_h^u(0), w)_{\mathscr{T}_h},$$

and the estimate follows readily.

Proposition 5.12 (An energy estimate) *For all $t \geqslant 0$*

$$\begin{aligned}
&\|\boldsymbol{\varepsilon}_h^q(t)\|_{\kappa^{-1}}^2 + |\mathrm{P}\varepsilon_h^u(t) - \widehat{\varepsilon}_h^u(t)|_\tau^2 + \|\dot{\varepsilon}_h^u(t)\|_\rho^2 \\
&\qquad \leqslant 2\|\mathbf{e}_h^q(0)\|_{\kappa^{-1}}^2 + 4\left(\int_0^t (\|\dot{\mathbf{e}}_h^q(t)\|_{\kappa^{-1}}^2 + \|\ddot{e}_h^u(t)\|_\rho^2)^{1/2} \mathrm{d}s\right)^2, \\
&\|\dot{\boldsymbol{\varepsilon}}_h^q(t)\|_{\kappa^{-1}}^2 + |\mathrm{P}\dot{\varepsilon}_h^u(t) - \dot{\widehat{\varepsilon}}_h^u(t)|_\tau^2 + \|\ddot{\varepsilon}_h^u(t)\|_\rho^2 \\
&\qquad \leqslant 2\|\dot{\mathbf{e}}_h^q(0)\|_{\kappa^{-1}}^2 + 2\|\ddot{e}_h^u(0)\|_\rho^2 + 4\left(\int_0^t (\|\ddot{\mathbf{e}}_h^q(t)\|_{\kappa^{-1}}^2 + \|\dddot{e}_h^u(t)\|_\rho^2)^{1/2} \mathrm{d}s\right)^2.
\end{aligned}$$

Proof Once again, the proof is reminiscent of what we did for similar estimates in the heat equation. We differentiate (5.39a) with respect to time and test it with $\mathbf{r} = \boldsymbol{\varepsilon}_h^q(t)$. We then test (5.39b) and (5.39c) with $w = \dot{\varepsilon}_h^u(t)$ and $\mu = \dot{\widehat{\varepsilon}}_h^u(t)$. The resulting equalities are added to produce the energy identity

$$\begin{aligned}
&\frac{1}{2}\frac{\mathrm{d}}{\mathrm{d}t}\left(\|\boldsymbol{\varepsilon}_h^q(t)\|_{\kappa^{-1}}^2 + |\mathrm{P}\varepsilon_h^u(t) - \widehat{\varepsilon}_h^u(t)|_\tau^2 + \|\dot{\varepsilon}_h^u(t)\|_\rho^2\right) \\
&\qquad\qquad = (\kappa^{-1}\dot{\mathbf{e}}_h^q(t), \boldsymbol{\varepsilon}_h^q(t))_{\mathscr{T}_h} + (\rho\ddot{e}_h^u(t), \dot{\varepsilon}_h^u(t))_{\mathscr{T}_h},
\end{aligned}$$

which is then integrated with respect to time

$$\begin{aligned}
&\|\boldsymbol{\varepsilon}_h^q(t)\|_{\kappa^{-1}}^2 + |\mathrm{P}\varepsilon_h^u(t) - \widehat{\varepsilon}_h^u(t)|_\tau^2 + \|\dot{\varepsilon}_h^u(t)\|_\rho^2 \\
&\quad = \|\boldsymbol{\varepsilon}_h^q(0)\|_{\kappa^{-1}}^2 + |\mathrm{P}\varepsilon_h^u(0) - \widehat{\varepsilon}_h^u(0)|_\tau^2 \\
&\qquad + 2\int_0^t \left((\kappa^{-1}\dot{\mathbf{e}}_h^q(s), \boldsymbol{\varepsilon}_h^q(s))_{\mathscr{T}_h} + (\rho\ddot{e}_h^u(s), \dot{\varepsilon}_h^u(s))_{\mathscr{T}_h}\right) \mathrm{d}s \\
&\quad \leqslant \|\mathbf{e}_h^q(0)\|_{\kappa^{-1}}^2 \\
&\qquad + \int_0^t 2\left(\|\dot{\mathbf{e}}_h^q(s)\|_{\kappa^{-1}}^2 + \|\ddot{e}_h^u(s)\|_\rho^2\right)^{1/2}\left(\|\boldsymbol{\varepsilon}_h^q(s)\|_{\kappa^{-1}}^2 + \|\dot{\varepsilon}_h^u(s)\|_\rho^2\right)^{1/2} \mathrm{d}s.
\end{aligned}$$

The above inequality fits in the framework of Lemma 5.1 with

$$\phi(t) := \left(\|\boldsymbol{\varepsilon}_h^q(t)\|_{\kappa^{-1}}^2 + |\mathrm{P}\varepsilon_h^u(t) - \widehat{\varepsilon}_h^u(t)|_\tau^2 + \|\dot{\varepsilon}_h^u(t)\|_\rho^2\right)^{1/2},$$
$$\gamma(t) := \|\mathbf{e}_h^q(0)\|_{\kappa^{-1}}^2,$$
$$\beta(t) := 2\left(\|\dot{\mathbf{e}}_h^q(t)\|_{\kappa^{-1}}^2 + \|\dddot{e}_h^u(t)\|_\rho^2\right)^{1/2},$$

and the first estimate of the proposition is just the thesis of that lemma. For the second one, differentiate the error equations with respect to time and follow a similar procedure for the proof.

5.3.2 *Estimates by Duality*

In the coming estimates, we will again be using the wiggled inequality symbol $\lesssim$, with the understanding that the hidden constant is independent not only of h but of the final time T. Dependence with respect to T will be always made explicit.

An identity involving the adjoint problem. As usual, we start with the solution of the dual-adjoint problem

$$\kappa^{-1}\boldsymbol{\xi}(t) - \nabla\theta(t) = \mathbf{0} \qquad \text{in } \Omega \quad t \in [0, T], \tag{5.40a}$$
$$-\operatorname{div}\boldsymbol{\xi}(t) + \rho\ddot{\theta}(t) = 0 \qquad \text{in } \Omega \quad t \in [0, T], \tag{5.40b}$$
$$\theta(t) = 0 \qquad \text{on } \Gamma \quad t \in [0, T], \tag{5.40c}$$
$$\theta(T) = 0, \tag{5.40d}$$
$$\dot{\theta}(T) = \varepsilon_h^u(T), \tag{5.40e}$$

its projection

$$(\boldsymbol{\Pi}\boldsymbol{\xi}(t), \Pi\theta(t)) := \Pi^{\mathrm{HDG+}}(\boldsymbol{\xi}(t), \theta(t); -\tau),$$

and the equations satisfied by the projection: for all $t \in [0, T]$ and $\mathbf{r} \in \mathbf{V}_h$, $w \in W_h$, $\mu \in M_h$,

$$(\kappa^{-1}\boldsymbol{\Pi}\boldsymbol{\xi}(t), \mathbf{r})_{\mathscr{T}_h} + (\Pi\theta(t), \operatorname{div}\mathbf{r})_{\mathscr{T}_h} - \langle \mathrm{P}\theta(t), \mathbf{r}\cdot\mathbf{n}\rangle_{\partial\mathscr{T}_h} = (\kappa^{-1}(\boldsymbol{\Pi}\boldsymbol{\xi}(t) - \boldsymbol{\xi}(t)), \mathbf{r})_{\mathscr{T}_h}, \tag{5.41a}$$
$$-(\operatorname{div}\boldsymbol{\Pi}\boldsymbol{\xi}(t), w)_{\mathscr{T}_h} + \langle\tau(\mathrm{P}\Pi\theta(t) - \theta(t)), w\rangle_{\partial\mathscr{T}_h} = -(\rho\ddot{\theta}(t), w)_{\mathscr{T}_h}, \tag{5.41b}$$
$$\langle \boldsymbol{\Pi}\boldsymbol{\xi}(t)\cdot\mathbf{n} - \tau(\mathrm{P}\Pi\theta(t) - \mathrm{P}\theta(t)), \mu\rangle_{\partial\mathscr{T}_h\setminus\Gamma} = 0, \tag{5.41c}$$
$$\langle \mathrm{P}\theta(t), \mu\rangle_\Gamma = 0. \tag{5.41d}$$

We then compare (5.41) tested with $(\boldsymbol{\varepsilon}_h^q(t), \varepsilon_h^u(t), \widehat{\varepsilon}_h^u(t))$ with the error equations tested with $(\boldsymbol{\Pi}\boldsymbol{\xi}(t), \Pi\theta(t), \mathrm{P}\theta(t))$ to obtain

$$
\begin{aligned}
(\kappa^{-1}(\boldsymbol{\Pi}\boldsymbol{\xi}(t)-\boldsymbol{\xi}(t)), \boldsymbol{\varepsilon}_h^q(t))_{\mathcal{T}_h} - (\rho\,\ddot\theta(t), \varepsilon_h^u(t))_{\mathcal{T}_h} \\
= (\rho(\ddot u_h(t) - \ddot u(t), \Pi\theta(t))_{\mathcal{T}_h} + (\kappa^{-1}\mathbf{e}_h^q(t), \boldsymbol{\Pi}\boldsymbol{\xi}(t))_{\mathcal{T}_h},
\end{aligned}
$$

or, equivalently,

$$
\begin{aligned}
(\rho\,\ddot\theta(t), \varepsilon_h^u(t))_{\mathcal{T}_h} =& (\kappa^{-1}(\boldsymbol{\Pi}\boldsymbol{\xi}(t) - \boldsymbol{\xi}(t)), \mathbf{q}(t) - \mathbf{q}_h(t))_{\mathcal{T}_h} - (\kappa^{-1}\boldsymbol{\xi}(t), \mathbf{e}_h^q(t))_{\mathcal{T}_h} \\
&+ (\rho(\ddot u(t) - \ddot u_h(t)), \Pi\theta(t))_{\mathcal{T}_h} \\
=& (\kappa^{-1}(\boldsymbol{\Pi}\boldsymbol{\xi}(t) - \boldsymbol{\xi}(t)), \mathbf{q}(t) - \mathbf{q}_h(t))_{\mathcal{T}_h} - (\kappa^{-1}\boldsymbol{\xi}(t), \mathbf{e}_h^q(t))_{\mathcal{T}_h} \\
&+ (\rho(\ddot u(t) - \ddot u_h(t)), \Pi\theta(t) - \theta(t))_{\mathcal{T}_h} \\
&+ (\rho(\ddot u(t) - \ddot u_h(t)), \theta(t))_{\mathcal{T}_h}.
\end{aligned}
$$

Consider the function

$$
\eta(t) := (\rho\,\dot\theta(t), \varepsilon_h^u(t))_{\mathcal{T}_h} - (\rho\,\theta(t), \dot\varepsilon_h^u(t))_{\mathcal{T}_h},
$$

and note that

$$
\eta(0) = (\rho\dot\theta(0), \varepsilon_h^u(0))_{\mathcal{T}_h}, \qquad \eta(T) = \|\varepsilon_h^u(T)\|_\rho^2,
$$

and (when $k \geqslant 1$)

$$
\begin{aligned}
\dot\eta(t) =& (\rho\,\ddot\theta(t), \varepsilon_h^u(t))_{\mathcal{T}_h} - (\rho\,\theta(t), \ddot\varepsilon_h^u(t))_{\mathcal{T}_h} \\
=& (\kappa^{-1}(\boldsymbol{\Pi}\boldsymbol{\xi}(t) - \boldsymbol{\xi}(t)), \mathbf{q}(t) - \mathbf{q}_h(t))_{\mathcal{T}_h} - (\nabla\theta(t), \mathbf{e}_h^q(t))_{\mathcal{T}_h} \\
&+ (\rho(\ddot u(t) - \ddot u_h(t)), \Pi\theta(t) - \theta(t))_{\mathcal{T}_h} - (\ddot e_h^u(t), \rho\,\theta(t))_{\mathcal{T}_h} \\
=& (\kappa^{-1}(\boldsymbol{\Pi}\boldsymbol{\xi}(t) - \boldsymbol{\xi}(t)), \mathbf{q}(t) - \mathbf{q}_h(t))_{\mathcal{T}_h} - (\nabla\theta(t) - \mathrm{P}_0\nabla\theta(t), \mathbf{e}_h^q(t))_{\mathcal{T}_h} \\
&+ (\rho(\ddot u(t) - \ddot u_h(t)), \Pi\theta(t) - \theta(t))_{\mathcal{T}_h} - (\ddot e_h^u(t), \rho\,\theta(t) - \mathrm{P}_0(\rho\theta(t)))_{\mathcal{T}_h}.
\end{aligned}
$$

This proves the identity (compare with Proposition 5.6 for the heat equation)

$$
\begin{aligned}
\|\varepsilon_h^u(T)\|_\rho^2 =& \int_0^T (\kappa^{-1}(\boldsymbol{\Pi}\boldsymbol{\xi}(t) - \boldsymbol{\xi}(t)), \mathbf{q}(t) - \mathbf{q}_h(t))_{\mathcal{T}_h} \mathrm{d}t && \text{(5.42a)} \\
&- \int_0^T (\nabla\theta(t) - \mathrm{P}_0\nabla\theta(t), \mathbf{e}_h^q(t))_{\mathcal{T}_h} \mathrm{d}t && \text{(5.42b)} \\
&+ \int_0^T (\rho(\ddot u(t) - \ddot u_h(t)), \Pi\theta(t) - \theta(t))_{\mathcal{T}_h} \mathrm{d}t && \text{(5.42c)} \\
&- \int_0^T (\ddot e_h^u(t), \rho\,\theta(t))_{\mathcal{T}_h} \mathrm{d}t && \text{(5.42d)} \\
&+ (\rho\dot\theta(0), \varepsilon_h^u(0))_{\mathcal{T}_h}. && \text{(5.42e)}
\end{aligned}
$$

We next estimate the terms in (5.42). The most complicated one is (5.42c), which will be handled by Proposition 5.13.

Regularity assumption and more notation. We assume the following elliptic regularity inequality

$$\|\kappa\nabla\theta\|_{1,\Omega} + \|\theta\|_{2,\Omega} \leqslant C\|\mathrm{div}\,(\kappa\nabla\theta)\|_{\Omega}, \tag{5.43}$$

holds for any $\theta \in H_0^1(\Omega)$ such that the right-hand side of the above equation is finite. We will use the notation for space–time norms introduced in (5.22). Furthermore, we define the norms

$$|\widehat{u}|_{\tau,p} := \left(\int_0^T |\widehat{u}(t)|_\tau^p \,\mathrm{d}t\right)^{1/p}.$$

The norm $|\cdot|_{\tau,\infty}$ is similarly defined with essential supremums over time instead of integrals in time. Finally, we define

$$\underline{\theta}(t) := \int_t^T \theta(s)ds, \quad \underline{\boldsymbol{\xi}}(t) := \int_t^T \boldsymbol{\xi}(s)ds.$$

Proposition 5.13 *Assuming (5.43), we have*

$$\left|\int_0^T (\rho(\ddot{u}(t) - \ddot{u}_h(t)), \Pi\theta(t) - \theta(t))_{\mathscr{T}_h} dt\right|$$
$$\lesssim h\|\varepsilon_h^u(T)\|_\Omega \Big(\|\boldsymbol{\varepsilon}_h^q(0)\|_\Omega + T\|\dot{\boldsymbol{\varepsilon}}_h^q\|_\infty + |\mathrm{P}\varepsilon_h^u(0) - \widehat{\varepsilon}_h^u(0)|_\tau$$
$$+ T|\mathrm{P}\dot{\varepsilon}_h^u - \dot{\widehat{\varepsilon}}_h^u|_{\tau,\infty} + h\|\ddot{e}_h^u(0)\|_\Omega + hT\|\dddot{e}_h^u\|_\infty \Big).$$

Proof In the coming arguments, and for the sake of shortening some estimates, we will prove bounds in terms of the quantity

$$\Xi(T) := \sup_{t\in[0,T]} |\underline{\theta}(t)|_{2,\Omega} + \sup_{t\in[0,T]} |\underline{\boldsymbol{\xi}}(t)|_{1,\Omega}. \tag{5.44}$$

We will prove in Proposition 5.16 that, assuming elliptic regularity, we have the estimate

$$\Xi(T) \lesssim \|\varepsilon_h^u(T)\|_\Omega. \tag{5.45}$$

Now let $e_h^\theta(t) = \Pi\theta(t) - \theta(t)$, and note that

$$\int_0^T (\rho(\ddot{u}(t) - \ddot{u}_h(t)), e_h^\theta(t))_{\mathscr{T}_h} \mathrm{d}t = \int_0^T (\rho(\ddot{\varepsilon}_h^u(t) - \ddot{e}_h^u(t)), e_h^\theta(t))_{\mathscr{T}_h} \mathrm{d}t. \tag{5.46}$$

For the second term of (5.46), we have

$$\left|\int_0^T (\rho \ddot{e}_h^u(t), e_h^\theta(t))_{\mathcal{T}_h} \mathrm{d}t\right| = \left|(\rho \ddot{e}_h^u(0), \underline{e}_h^\theta(0))_{\mathcal{T}_h} + \int_0^T (\rho \dddot{e}_h^u(t), \underline{e}_h^\theta(t))_{\mathcal{T}_h} \mathrm{d}t\right|$$
$$\lesssim h^2(\|\ddot{e}_h^u(0)\|_\Omega + T\|\dddot{e}_h^u\|_\infty)\Xi(T).$$

We next estimate the remaining term in (5.46). Since $\ddot{\varepsilon}_h^u(t)\big|_K \in \mathcal{P}_{k+1}(K)$ for all K, we have

$$\int_0^T (\rho \ddot{\varepsilon}_h^u(t), e_h^\theta(t))_{\mathcal{T}_h} \mathrm{d}t = \int_0^T (\rho \ddot{\varepsilon}_h^u(t), \mathrm{P}_{k+1}^\rho e_h^\theta(t))_{\mathcal{T}_h} \mathrm{d}t,$$

where, P_{k+1}^ρ is the ρ-weighted L^2 projection onto $\mathcal{P}_{k+1}(K)$. Testing the second error Eq. (5.39b) with $w = \mathrm{P}_{k+1}^\rho e_h^\theta$, we have

$$\begin{aligned}(\rho \ddot{\varepsilon}_h^u(t), \mathrm{P}_{k+1}^\rho e_h^\theta(t))_{\mathcal{T}_h} &= -(\operatorname{div} \boldsymbol{\varepsilon}_h^q(t), \mathrm{P}_{k+1}^\rho e_h^\theta(t))_{\mathcal{T}_h} \\ &\quad - \langle \tau \mathrm{P}(\varepsilon_h^u(t) - \widehat{\varepsilon}_h^u(t)), \mathrm{P}_{k+1}^\rho e_h^\theta(t)\rangle_{\partial\mathcal{T}_h} \\ &\quad + (\rho \ddot{e}_h^u(t), \mathrm{P}_{k+1}^\rho e_h^\theta(t))_{\mathcal{T}_h} \\ &=: Q_1(t) + Q_2(t) + Q_3(t).\end{aligned}$$

Now we use integration by parts in time to estimate the above three terms:

$$\begin{aligned}\int_0^T Q_1(t)\mathrm{d}t &= -(\operatorname{div} \boldsymbol{\varepsilon}_h^q(0), \mathrm{P}_{k+1}^\rho \underline{e}_h^\theta(0))_{\mathcal{T}_h} - \int_0^T (\operatorname{div} \dot{\boldsymbol{\varepsilon}}_h^q(t), \mathrm{P}_{k+1}^\rho \underline{e}_h^\theta(t))_{\mathcal{T}_h} \mathrm{d}t, \\ \int_0^T Q_2(t)\mathrm{d}t &= -\langle \tau \mathrm{P}(\varepsilon_h^u(0) - \widehat{\varepsilon}_h^u(0)), \mathrm{P}_{k+1}^\rho \underline{e}_h^\theta(0)\rangle_{\partial\mathcal{T}_h} \\ &\quad - \int_0^T \langle \tau \mathrm{P}(\dot{\varepsilon}_h^u(t) - \dot{\widehat{\varepsilon}}_h^u(t)), \mathrm{P}_{k+1}^\rho \underline{e}_h^\theta(t)\rangle_{\partial\mathcal{T}_h} \mathrm{d}t, \\ \int_0^T Q_3(t)\mathrm{d}t &= (\rho \ddot{e}_h^u(0), \mathrm{P}_{k+1}^\rho \underline{e}_h^\theta(0))_{\mathcal{T}_h} + \int_0^T (\rho \dddot{e}_h^u(t), \mathrm{P}_{k+1}^\rho \underline{e}_h^\theta(t))_{\mathcal{T}_h} \mathrm{d}t.\end{aligned}$$

Note that

$$\mathrm{P}_{k+1}^\rho \underline{e}_h^\theta(t) = \mathrm{P}_{k+1}^\rho (\Pi \underline{\theta}(t) - \underline{\theta}(t)) = \Pi \underline{\theta}(t) - \underline{\theta}(t) - \mathrm{P}_{k+1}^\rho \underline{\theta}(t) + \underline{\theta}(t).$$

Combining the above with the convergence properties about $\boldsymbol{\Pi}\boldsymbol{\xi}(t)$ and $\Pi\theta(t)$ (see Proposition 4.6), we have

$$\begin{aligned}\|\mathrm{P}_{k+1}^\rho \underline{e}_h^\theta(t)\|_K &\lesssim h_K^2(|\underline{\theta}(t)|_{2,K} + |\boldsymbol{\xi}(t)|_{1,K}), \\ \|\tau^{1/2}\mathrm{P}_{k+1}^\rho \underline{e}_h^\theta(t)\|_{\partial K} &\lesssim h_K(|\underline{\theta}(t)|_{2,K} + |\boldsymbol{\xi}(t)|_{1,K}).\end{aligned}$$

Now back to the estimate of $Q_i(t)$, we have

$$\left|\int_0^T Q_1(t)\mathrm{d}t\right| \lesssim h(\|\boldsymbol{\varepsilon}_h^q(0)\|_\Omega + T\|\dot{\boldsymbol{\varepsilon}}_h^q\|_\infty)\Xi(T),$$
$$\left|\int_0^T Q_2(t)\mathrm{d}t\right| \lesssim h(|\mathrm{P}\varepsilon_h^u(0) - \widehat{\varepsilon}_h^u(0)|_\tau + T|\mathrm{P}\dot{\varepsilon}_h^u - \dot{\widehat{\varepsilon}}_h^u|_{\tau,\infty})\Xi(T),$$
$$\left|\int_0^T Q_3(t)\mathrm{d}t\right| \lesssim h^2(\|\ddot{e}_h^u(0)\|_\Omega + T\|\dddot{e}_h^u\|_\infty)\Xi(T),$$

where we used the fact that $\|\mathrm{div}\,\mathbf{q}\|_K \lesssim h_K^{-1}\|\mathbf{q}\|_K$ for any $\mathbf{q} \in \mathscr{P}_k(K)$ (to see this fact, use (1.9) and finite-dimensional arguments). Finally, we use (5.45) to bound $\Xi(T)$ and the proof is finished.

Proposition 5.14 *Assuming $k \geqslant 1$ and (5.43), we have*

$$\|\varepsilon_h^u(T)\|_\Omega \lesssim (1+T)^2\left(\|\ddot{e}_h^u\|_\infty + h\|\dddot{e}_h^u\|_\infty + h\|\mathbf{e}_h^q(0)\|_\Omega + h\|\dot{\mathbf{e}}_h^q\|_\infty + h\|\ddot{\mathbf{e}}_h^q\|_\infty\right).$$

Proof We will estimate those terms in (5.42). Proposition 5.15 allows us to handle (5.42d):

$$\left|\int_0^T (\ddot{e}_h^u(t), \rho\theta(t))_{\mathscr{T}_h} dt\right| \lesssim \|\ddot{e}_h^u\|_1\|\theta\|_\infty \leqslant T\|\ddot{e}_h^u\|_\infty\|\varepsilon_h^u(T)\|_\Omega.$$

An estimate for (5.42b) is obtained as follows:

$$\begin{aligned}
&\left|\int_0^T (\mathrm{P}_0\nabla\theta(t) - \nabla\theta(t), \mathbf{e}_h^q(t))_{\mathscr{T}_h} dt\right| \\
&= \left|(\mathrm{P}_0\nabla\underline{\theta}(0) - \nabla\underline{\theta}(0), \mathbf{e}_h^q(0))_{\mathscr{T}_h} + \int_0^T (\mathrm{P}_0\nabla\underline{\theta}(t) - \nabla\underline{\theta}(t), \dot{\mathbf{e}}_h^q(t))_{\mathscr{T}_h}\mathrm{d}t\right| \\
&\lesssim h \sup_{t\in[0,T]} |\underline{\theta}(t)|_{2,\Omega}(\|\mathbf{e}_h^q(0)\|_\Omega + \|\dot{\mathbf{e}}_h^q\|_1) \\
&\lesssim h\|\varepsilon_h^u(T)\|_\Omega(1+T)(\|\mathbf{e}_h^q(0)\|_\Omega + \|\dot{\mathbf{e}}_h^q\|_\infty). \qquad \text{(by (5.45))}
\end{aligned}$$

For the term (5.42a), we have

$$\begin{aligned}
&\left|\int_0^T (\boldsymbol{\Pi}\boldsymbol{\xi}(t) - \boldsymbol{\xi}(t), \kappa^{-1}(\mathbf{q}(t) - \mathbf{q}_h(t)))_{\mathscr{T}_h}\mathrm{d}t\right| \\
&\leqslant \left|(\boldsymbol{\Pi}\underline{\boldsymbol{\xi}}(0) - \underline{\boldsymbol{\xi}}(0), \kappa^{-1}(\mathbf{q}(0) - \mathbf{q}_h(0)))_{\mathscr{T}_h}\right| \\
&\quad + \left|\int_0^T (\boldsymbol{\Pi}\underline{\boldsymbol{\xi}}(t) - \underline{\boldsymbol{\xi}}(t), \kappa^{-1}(\dot{\mathbf{q}}(t) - \dot{\mathbf{q}}_h(t)))_{\mathscr{T}_h}\mathrm{d}t\right| \\
&\lesssim h\Xi(T)\left(\|\mathbf{q}(0) - \mathbf{q}_h(0)\|_\Omega + \|\dot{\mathbf{q}} - \dot{\mathbf{q}}_h\|_1\right) \\
&\lesssim h\|\varepsilon_h^u(T)\|_\Omega\left(\|\boldsymbol{\varepsilon}_h^q(0)\|_\Omega + \|\mathbf{e}_h^q(0)\|_\Omega + T(\|\dot{\boldsymbol{\varepsilon}}_h^q\|_\infty + \|\dot{\mathbf{e}}_h^q\|_\infty)\right),
\end{aligned}$$

where we used (5.45) again in the last step of the above estimates. For the last term (5.42e), we have

$$|(\rho\dot{\theta}(0), \varepsilon_h^u(0))_{\mathcal{T}_h}| \lesssim \|\dot{\theta}\|_\infty \|\varepsilon_h^u(0)\|_\Omega \lesssim h\|\varepsilon_h^u(T)\|_\Omega \|\mathbf{e}_h^q(0)\|_\Omega,$$

where we use Proposition 5.15 to bound $\|\dot{\theta}\|_\infty$ and (4.18) to estimate $\varepsilon_h^u(0)$ (error for a steady-state system).

Now combining Eqs. (5.42), the above four estimates, and Proposition 5.13, we have

$$\begin{aligned}\|\varepsilon_h^u(T)\|_\Omega \lesssim & T\|\ddot{e}_h^u\|_\infty + h(1+T)(\|\mathbf{e}_h^q(0)\|_\Omega + \|\dot{\mathbf{e}}_h^q\|_\infty)\\ &+ h\left(\|\boldsymbol{\varepsilon}_h^q(0)\|_\Omega + T\|\dot{\boldsymbol{\varepsilon}}_h^q\|_\infty\right)\\ &+ h\left(|\mathrm{P}\varepsilon_h^u(0) - \widehat{\varepsilon}_h^u(0)|_\tau + T|\mathrm{P}\dot{\varepsilon}_h^u - \dot{\widehat{\varepsilon}}_h^u|_{\tau,\infty} + h\|\ddot{e}_h^u(0)\|_\Omega + hT\|\dddot{e}_h^u\|_\infty\right).\end{aligned}$$

The rest of the proof follows by using Propositions 5.11 and 5.12, which imply that

$$\begin{aligned}\|\boldsymbol{\varepsilon}_h^q(0)\|_\Omega + |\mathrm{P}\varepsilon_h^u(0) - \widehat{\varepsilon}_h^u(0)|_\tau &\lesssim \|\mathbf{e}_h^q(0)\|_\Omega,\\ \|\dot{\boldsymbol{\varepsilon}}_h^q\|_\infty + |\mathrm{P}\dot{\varepsilon}_h^u - \dot{\widehat{\varepsilon}}_h^u|_{\tau,\infty} &\lesssim \|\dot{\mathbf{e}}_h^q(0)\|_\Omega + \|\ddot{e}_h^u(0)\|_\Omega + T\|\ddot{\mathbf{e}}_h^q\|_\infty + T\|\dddot{e}_h^u\|_\infty.\end{aligned}$$

5.3.3 *Regularity Estimates for the Wave Equation*

Here we give some known estimates for the solutions of the wave equation. Similar results can be found in [54].

Proposition 5.15 *For the dual Eqs. (5.40), we have*

$$\|\theta\|_{L^\infty(0,T;H^1(\Omega))} + \|\dot{\theta}\|_{L^\infty(0,T;L^2(\Omega))} \lesssim \|\varepsilon_h^u(T)\|_\Omega.$$

Proof Note first that $\operatorname{div}(\kappa\nabla\theta(t)) = \rho\ddot{\theta}(t)$, and therefore

$$\frac{\mathrm{d}}{\mathrm{d}t}\left((\rho\dot{\theta}(t), \dot{\theta}(t))_\Omega + (\kappa\nabla\theta(t), \nabla\theta(t))_\Omega\right) = 0 \quad \text{for } t \in [0, T].$$

Noticing that $\theta(T) = 0$ and $\dot{\theta}(T) = \varepsilon_h^u(T)$, we have

$$(\rho\dot{\theta}(t), \dot{\theta}(t))_\Omega + (\kappa\nabla\theta(t), \nabla\theta(t))_\Omega = (\rho\varepsilon_h^u(T), \varepsilon_h^u(T))_\Omega \quad \text{for } t \in [0, T].$$

The rest of the proof follows by using the boundary condition $\theta(t)\big|_\Gamma = 0$ (for all $t \geqslant 0$) and the Poincaré–Friedrichs inequality.

A combination of the above proposition and the regularity assumption (5.43) gives the following proposition.

Proposition 5.16 *Assume (5.43), then we have*

$$\|\underline{\boldsymbol{\xi}}\|_{L^\infty(0,T;H^1(\Omega))} + \|\underline{\theta}\|_{L^\infty(0,T;H^2(\Omega))} \lesssim \|\varepsilon_h^u(T)\|_\Omega.$$

Proof Integrating $\operatorname{div}(\kappa\nabla\theta)(t) = \rho\ddot{\theta}(t)$from t to T yields

$$\operatorname{div}(\kappa\nabla\underline{\theta}(t)) = \rho(\dot{\theta}(T) - \dot{\theta}(t)).$$

Therefore, for any $t \in [0, T]$, by (5.43) and Proposition 5.15, we have

$$\|\kappa\nabla\underline{\theta}(t)\|_{1,\Omega} + \|\underline{\theta}(t)\|_{2,\Omega} \lesssim \|\operatorname{div}(\kappa\nabla\underline{\theta}(t))\|_\Omega \lesssim \|\dot{\theta}(t)\|_\Omega + \|\dot{\theta}(T)\|_\Omega \lesssim \|\varepsilon_h^u(T)\|_\Omega.$$

Recalling that $\underline{\boldsymbol{\xi}}(t) = \kappa\nabla\underline{\theta}(t)$ completes the proof.

Exercises

1. Using the HDG semidiscrete Eqs. (5.32), show that if $f \equiv 0$ and $g \equiv 0$, the energy

$$\frac{1}{2}\|\dot{u}_h(t)\|_\rho^2 + \frac{1}{2}\|\mathbf{q}_h(t)\|_{\kappa^{-1}}^2 + \frac{1}{2}|\mathrm{P}u_h(t) - \widehat{u}_h(t)|_\tau^2$$

is constant over time.

Chapter 6
Further Reading

The exposition given in this monograph is limited to triangular–tetrahedral partitions of polygons–polyhedra, but much of it can be easily extended to other partitions, using new tricks to choose polynomial spaces and projections. We have also limited ourselves to linear steady-state diffusion, Helmholtz, heat, and wave equations, but the applications of HDG go much further into nonlinear, vector-valued, or non-selfadjoint equations. In this very short chapter, we will give some hints at "classics" or relevant recent contributions to the analysis and practice of HDG methods.

The Hybridizable Discontinuous Galerkin methods crystallized from the Local Discontinuous Galerkin methods in a series of papers that directly pointed out at a large but well-defined framework of discretization methods that included traditional mixed methods as a particular case and was the basis for the creation of many new very efficient methods. For those who might be skeptical about whether HDG competes with traditional FEM methods, let us point at two advantages: (a) since they are mixed methods in their core, HDG methods approximate well the main unknown and its associated flux, and tend to be locking free; (b) the bookkeeping (counting local and global degrees of freedom) for high-order methods is much simpler in HDG than in FEM [74]. This goes along with some frequently repeated claims about the pros of DG methods: hanging nodes, variable degrees, etc. Some early references showing how HDG was being constructed and analyzed are [30, 51], with [43] acting as the presentation in society of HDG and [45] introducing the projection-based analysis, which is where this monograph puts its emphasis.

To cite applications to other equations, let us just refer to articles on

- Elasticity [61, 70, 83, 84, 114, 116],
- The Helmholtz equation [80, 81, 100],
- Convection–diffusion problems and linear equations of third order [19–21, 32, 66, 73, 95, 101, 111],
- Eigenvalues of elliptic operators [79],
- The Stokes equation, which admits many different first-order formulations, by choosing different unknowns, and the associated Brinkman equation [26, 27, 42, 44, 48, 57, 60, 62, 71, 72, 76–78, 96],

S. Du and F.-J. Sayas, *An Invitation to the Theory of the Hybridizable Discontinuous Galerkin Method*, SpringerBriefs in Mathematics,
https://doi.org/10.1007/978-3-030-27230-2_6

- The Navier–Stokes equation [13, 14, 38, 86, 88, 94, 99, 105, 110],
- Maxwell equations [17, 18, 89, 98],
- Some evolutionary equations (heat, wave, fractional) [15, 40, 47, 54, 82, 91, 97, 112, 117],
- Coupling with Discontinuous Petrov–Galerkin methods and Boundary Element Methods [46, 56, 58, 75, 90],
- A posteriori error estimates [16, 50, 64, 65],
- Treatment of curved boundaries [52, 53],
- Nonlinear equations [59, 63, 84, 119, 120], and
- A variety of problems arising from beam-plate equations or PDE on surfaces [11, 12, 28, 31].

What we have called here the HDG+ method stems from the idea of projecting higher order polynomial approximations in the interior of the tetrahedra into a lower order polynomial space on the boundary. This appears in the context of a somewhat different HDG scheme, based on second-order-in-space formulation and is due to Lehrenfeld and Schöberl. Some works on HDG+ methods, which seem to be highly relevant for optimally convergent methods in elasticity with symmetric approximation of the stress tensor, are [18, 87, 102–104, 106–108]. A considerable effort has been spent on comparing and relating HDG methods to other new similar families of methods (they tend to have coincidences when taking limits in some parameters, or for the lowest order classes) like the Staggered Discontinuous Galerkin method (SDG) or the Hybrid High Order (HHO) scheme of DiPietro and Ern [22, 23, 29].

Practical comparisons with FEM and implementation techniques in Matlab can be found in [74, 85, 121]. Solvers (multigrid, domain decomposition) and multiscale techniques are explored in [33, 67, 68, 113]. Two recent surveys by some of the highest bidders in the HDG community [25, 49] give a precise account of the development of this family of methods.

We finally mention a recent expansion, due to Cockburn, Fu, Qiu, and Sayas, on the theory and construction of HDG methods called M-decompositions [34–37, 39]. The idea behind this is to construct triples of spaces $(\mathbf{V}_h, W_h, M_h)$ so that the associated HDG method is optimally convergent (and superconvergent in the main variable). The goal is to handle general polyhedral partitions and to give a systematic form to build these triples. This is done by starting with a basic polynomial triple and enrich it until something like Lemma 2.2 holds (the space on the left of (2.3) acts as the local M-space, which needs to be decomposed as two different trace spaces, hence the name of M-decomposition). The idea has been fundamental to derive new families of methods on parallelepipeds and more general polyhedra. While successful, the constructions derived with this abstract methodology are not always simple, and the resulting approximation spaces can contain complicated functions. On the other hand, the M-decompositions have been successfully used by the authors of this monograph to find HDG+ projections for elasticity and electromagnetism, without having to use the M-decomposition spaces in practice.

References

1. Arnold, D.N., Brezzi, F.: Mixed and nonconforming finite element methods: implementation, postprocessing and error estimates. RAIRO Modél. Math. Anal. Numér. **19**(1), 7–32 (1985)
2. Arnold, D.N., Brezzi, F., Cockburn, B., Marini, L.D.: Unified analysis of discontinuous Galerkin methods for elliptic problems. SIAM J. Numer. Anal. **39**(5), 1749–1779 (2001/02). https://doi.org/10.1137/S0036142901384162
3. Arnold, D.N., Falk, R.S., Winther, R.: Finite element exterior calculus, homological techniques, and applications. Acta Numer. **15**, 1–155 (2006). https://doi.org/10.1017/S0962492906210018
4. Arnold, D.N., Falk, R.S., Winther, R.: Finite element exterior calculus: from Hodge theory to numerical stability. Bull. Am. Math. Soc. (N.S.) **47**(2), 281–354 (2010). https://doi.org/10.1090/S0273-0979-10-01278-4
5. Braess, D.: Finite Elements, 3rd edn. Cambridge University Press, Cambridge (2007). https://doi.org/10.1017/CBO9780511618635; Theory, fast solvers, and applications in elasticity theory, Translated from the German by Larry L. Schumaker
6. Brenner, S.C., Scott, L.R.: The mathematical theory of finite element methods. Texts in Applied Mathematics, vol. 15, 3rd edn. Springer, New York (2008). https://doi.org/10.1007/978-0-387-75934-0
7. Brezzi, F.: On the existence, uniqueness and approximation of saddle-point problems arising from Lagrangian multipliers. Rev. Française Automat. Informat. Recherche Opérationnelle Sér. Rouge **8**(R-2), 129–151 (1974)
8. Brezzi, F., Douglas Jr., J., Durán, R., Fortin, M.: Mixed finite elements for second order elliptic problems in three variables. Numer. Math. **51**(2), 237–250 (1987). https://doi.org/10.1007/BF01396752
9. Brezzi, F., Douglas Jr., J., Marini, L.D.: Two families of mixed finite elements for second order elliptic problems. Numer. Math. **47**(2), 217–235 (1985). https://doi.org/10.1007/BF01389710
10. Brezzi, F., Fortin, M.: Mixed and hybrid finite element methods. Springer Series in Computational Mathematics, vol. 15. Springer, New York (1991). https://doi.org/10.1007/978-1-4612-3172-1
11. Celiker, F., Cockburn, B., Shi, K.: Hybridizable discontinuous Galerkin methods for Timoshenko beams. J. Sci. Comput. **44**(1), 1–37 (2010). https://doi.org/10.1007/s10915-010-9357-2
12. Celiker, F., Cockburn, B., Shi, K.: A projection-based error analysis of HDG methods for Timoshenko beams. Math. Comput. **81**(277), 131–151 (2012). https://doi.org/10.1090/S0025-5718-2011-02522-6
13. Cesmelioglu, A., Cockburn, B., Nguyen, N.C., Peraire, J.: Analysis of HDG methods for Oseen equations. J. Sci. Comput. **55**(2), 392–431 (2013). https://doi.org/10.1007/s10915-012-9639-y

S. Du and F.-J. Sayas, *An Invitation to the Theory of the Hybridizable Discontinuous Galerkin Method*, SpringerBriefs in Mathematics,
https://doi.org/10.1007/978-3-030-27230-2

14. Cesmelioglu, A., Cockburn, B., Qiu, W.: Analysis of a hybridizable discontinuous Galerkin method for the steady-state incompressible Navier-Stokes equations. Math. Comput. **86**(306), 1643–1670 (2017). https://doi.org/10.1090/mcom/3195
15. Chabaud, B., Cockburn, B.: Uniform-in-time superconvergence of HDG methods for the heat equation. Math. Comput. **81**(277), 107–129 (2012). https://doi.org/10.1090/S0025-5718-2011-02525-1
16. Chen, H., Li, J., Qiu, W.: Robust a posteriori error estimates for HDG method for convection-diffusion equations. IMA J. Numer. Anal. **36**(1), 437–462 (2016). https://doi.org/10.1093/imanum/drv009
17. Chen, H., Qiu, W., Shi, K.: A priori and computable a posteriori error estimates for an HDG method for the coercive Maxwell equations. Comput. Methods Appl. Mech. Eng. **333**, 287–310 (2018). https://doi.org/10.1016/j.cma.2018.01.030
18. Chen, H., Qiu, W., Shi, K., Solano, M.: A superconvergent HDG method for the Maxwell equations. J. Sci. Comput. **70**(3), 1010–1029 (2017). https://doi.org/10.1007/s10915-016-0272-z
19. Chen, Y., Cockburn, B.: Analysis of variable-degree HDG methods for convection-diffusion equations. Part I: general nonconforming meshes. IMA J. Numer. Anal. **32**(4), 1267–1293 (2012). https://doi.org/10.1093/imanum/drr058
20. Chen, Y., Cockburn, B.: Analysis of variable-degree HDG methods for convection-diffusion equations. Part II: semimatching nonconforming meshes. Math. Comput. **83**(285), 87–111 (2014). https://doi.org/10.1090/S0025-5718-2013-02711-1
21. Chen, Y., Cockburn, B., Dong, B.: Superconvergent HDG methods for linear, stationary, third-order equations in one-space dimension. Math. Comput. **85**(302), 2715–2742 (2016). https://doi.org/10.1090/mcom/3091
22. Chung, E., Cockburn, B., Fu, G.: The staggered DG method is the limit of a hybridizable DG method. SIAM J. Numer. Anal. **52**(2), 915–932 (2014). https://doi.org/10.1137/13091573X
23. Chung, E., Cockburn, B., Fu, G.: The staggered DG method is the limit of a hybridizable DG method. Part II: the Stokes flow. J. Sci. Comput. **66**(2), 870–887 (2016). https://doi.org/10.1007/s10915-015-0047-y
24. Ciarlet, P.G.: The Finite Element Method for Elliptic Problems. Studies in Mathematics and its Applications, vol. 4. North-Holland Publishing Co., Amsterdam (1978)
25. Cockburn, B.: Static condensation, hybridization, and the devising of the HDG methods. Building Bridges: Connections and Challenges in Modern Approaches to Numerical Partial Differential Equations. Lecture Notes of Computer Science & Engineering, vol. 114, pp. 129–177. Springer, Cham (2016)
26. Cockburn, B., Cui, J.: An analysis of HDG methods for the vorticity-velocity-pressure formulation of the Stokes problem in three dimensions. Math. Comput. **81**(279), 1355–1368 (2012). https://doi.org/10.1090/S0025-5718-2011-02575-5
27. Cockburn, B., Cui, J.: Divergence-free HDG methods for the vorticity-velocity formulation of the Stokes problem. J. Sci. Comput. **52**(1), 256–270 (2012). https://doi.org/10.1007/s10915-011-9542-y
28. Cockburn, B., Demlow, A.: Hybridizable discontinuous Galerkin and mixed finite element methods for elliptic problems on surfaces. Math. Comput. **85**(302), 2609–2638 (2016). https://doi.org/10.1090/mcom/3093
29. Cockburn, B., Di Pietro, D.A., Ern, A.: Bridging the hybrid high-order and hybridizable discontinuous Galerkin methods. ESAIM Math. Model. Numer. Anal. **50**(3), 635–650 (2016). https://doi.org/10.1051/m2an/2015051
30. Cockburn, B., Dong, B., Guzmán, J.: A superconvergent LDG-hybridizable Galerkin method for second-order elliptic problems. Math. Comput. **77**(264), 1887–1916 (2008). https://doi.org/10.1090/S0025-5718-08-02123-6
31. Cockburn, B., Dong, B., Guzmán, J.: A hybridizable and superconvergent discontinuous Galerkin method for biharmonic problems. J. Sci. Comput. **40**(1–3), 141–187 (2009). https://doi.org/10.1007/s10915-009-9279-z
32. Cockburn, B., Dong, B., Guzmán, J., Restelli, M., Sacco, R.: A hybridizable discontinuous Galerkin method for steady-state convection-diffusion-reaction problems. SIAM J. Sci. Comput. **31**(5), 3827–3846 (2009). https://doi.org/10.1137/080728810

33. Cockburn, B., Dubois, O., Gopalakrishnan, J., Tan, S.: Multigrid for an HDG method. IMA J. Numer. Anal. **34**(4), 1386–1425 (2014). https://doi.org/10.1093/imanum/drt024
34. Cockburn, B., Fu, G.: Superconvergence by M-decompositions. Part II: construction of two-dimensional finite elements. ESAIM Math. Model. Numer. Anal. **51**(1), 165–186 (2017). https://doi.org/10.1051/m2an/2016016
35. Cockburn, B., Fu, G.: Superconvergence by M-decompositions. Part III: Construction of three-dimensional finite elements. ESAIM Math. Model. Numer. Anal. **51**(1), 365–398 (2017). https://doi.org/10.1051/m2an/2016023
36. Cockburn, B., Fu, G.: Devising superconvergent HDG methods with symmetric approximate stresses for linear elasticity by M-decompositions. IMA J. Numer. Anal. **38**(2), 566–604 (2018). https://doi.org/10.1093/imanum/drx025
37. Cockburn, B., Fu, G., Qiu, W.: A note on the devising of superconvergent HDG methods for Stokes flow by M-decompositions. IMA J. Numer. Anal. **37**(2), 730–749 (2017). https://doi.org/10.1093/imanum/drw029
38. Cockburn, B., Fu, G., Qiu, W.: Discrete H^1-inequalities for spaces admitting M-decompositions. SIAM J. Numer. Anal. **56**(6), 3407–3429 (2018). https://doi.org/10.1137/17M1144830
39. Cockburn, B., Fu, G., Sayas, F.J.: Superconvergence by M-decompositions. Part I: general theory for HDG methods for diffusion. Math. Comput. **86**(306), 1609–1641 (2017). https://doi.org/10.1090/mcom/3140
40. Cockburn, B., Fu, Z., Hungria, A., Ji, L., Sánchez, M.A., Sayas, F.J.: Stormer-Numerov HDG methods for acoustic waves. J. Sci. Comput. **75**(2), 597–624 (2018). https://doi.org/10.1007/s10915-017-0547-z
41. Cockburn, B., Gopalakrishnan, J.: A characterization of hybridized mixed methods for second order elliptic problems. SIAM J. Numer. Anal. **42**(1), 283–301 (electronic) (2004). https://doi.org/10.1137/S0036142902417893
42. Cockburn, B., Gopalakrishnan, J.: The derivation of hybridizable discontinuous Galerkin methods for Stokes flow. SIAM J. Numer. Anal. **47**(2), 1092–1125 (2009). https://doi.org/10.1137/080726653
43. Cockburn, B., Gopalakrishnan, J., Lazarov, R.: Unified hybridization of discontinuous Galerkin, mixed, and continuous Galerkin methods for second order elliptic problems. SIAM J. Numer. Anal. **47**(2), 1319–1365 (2009). https://doi.org/10.1137/070706616
44. Cockburn, B., Gopalakrishnan, J., Nguyen, N.C., Peraire, J., Sayas, F.J.: Analysis of HDG methods for Stokes flow. Math. Comput. **80**(274), 723–760 (2011). https://doi.org/10.1090/S0025-5718-2010-02410-X
45. Cockburn, B., Gopalakrishnan, J., Sayas, F.J.: A projection-based error analysis of HDG methods. Math. Comput. **79**(271), 1351–1367 (2010). https://doi.org/10.1090/S0025-5718-10-02334-3
46. Cockburn, B., Guzmán, J., Sayas, F.J.: Coupling of Raviart-Thomas and hybridizable discontinuous Galerkin methods with BEM. SIAM J. Numer. Anal. **50**(5), 2778–2801 (2012). https://doi.org/10.1137/100818339
47. Cockburn, B., Mustapha, K.: A hybridizable discontinuous Galerkin method for fractional diffusion problems. Numer. Math. **130**(2), 293–314 (2015). https://doi.org/10.1007/s00211-014-0661-x
48. Cockburn, B., Nguyen, N.C., Peraire, J.: A comparison of HDG methods for Stokes flow. J. Sci. Comput. **45**(1–3), 215–237 (2010). https://doi.org/10.1007/s10915-010-9359-0
49. Cockburn, B., Nguyen, N.C., Peraire, J.: HDG methods for hyperbolic problems. In: Handbook of Numerical Methods for Hyperbolic Problems. Handbook of Numerical Analysis, vol. 17, pp. 173–197. Elsevier/North-Holland, Amsterdam (2016)
50. Cockburn, B., Nochetto, R.H., Zhang, W.: Contraction property of adaptive hybridizable discontinuous Galerkin methods. Math. Comput. **85**(299), 1113–1141 (2016). https://doi.org/10.1090/mcom/3014

51. Cockburn, B., Qiu, W., Shi, K.: Conditions for superconvergence of HDG methods for second-order elliptic problems. Math. Comput. **81**(279), 1327–1353 (2012). https://doi.org/10.1090/S0025-5718-2011-02550-0
52. Cockburn, B., Qiu, W., Shi, K.: Superconvergent HDG methods on isoparametric elements for second-order elliptic problems. SIAM J. Numer. Anal. **50**(3), 1417–1432 (2012). https://doi.org/10.1137/110840790
53. Cockburn, B., Qiu, W., Solano, M.: A priori error analysis for HDG methods using extensions from subdomains to achieve boundary conformity. Math. Comput. **83**(286), 665–699 (2014). https://doi.org/10.1090/S0025-5718-2013-02747-0
54. Cockburn, B., Quenneville-Bélair, V.: Uniform-in-time superconvergence of the HDG methods for the acoustic wave equation. Math. Comput. **83**(285), 65–85 (2014). https://doi.org/10.1090/S0025-5718-2013-02743-3
55. Cockburn, B., Sánchez, M.A., Xiong, C.: Supercloseness of primal-dual Galerkin approximations for second order elliptic problems. J. Sci. Comput. **75**(1), 376–394 (2018). https://doi.org/10.1007/s10915-017-0538-0
56. Cockburn, B., Sayas, F.J.: The devising of symmetric couplings of boundary element and discontinuous Galerkin methods. IMA J. Numer. Anal. **32**(3), 765–794 (2012). https://doi.org/10.1093/imanum/drr019
57. Cockburn, B., Sayas, F.J.: Divergence-conforming HDG methods for Stokes flows. Math. Comput. **83**(288), 1571–1598 (2014). https://doi.org/10.1090/S0025-5718-2014-02802-0
58. Cockburn, B., Sayas, F.J., Solano, M.: Coupling at a distance HDG and BEM. SIAM J. Sci. Comput. **34**(1), A28–A47 (2012). https://doi.org/10.1137/110823237
59. Cockburn, B., Shen, J.: A hybridizable discontinuous Galerkin method for the p-Laplacian. SIAM J. Sci. Comput. **38**(1), A545–A566 (2016). https://doi.org/10.1137/15M1008014
60. Cockburn, B., Shi, K.: Conditions for superconvergence of HDG methods for Stokes flow. Math. Comput. **82**(282), 651–671 (2013). https://doi.org/10.1090/S0025-5718-2012-02644-5
61. Cockburn, B., Shi, K.: Superconvergent HDG methods for linear elasticity with weakly symmetric stresses. IMA J. Numer. Anal. **33**(3), 747–770 (2013). https://doi.org/10.1093/imanum/drs020
62. Cockburn, B., Shi, K.: Devising HDG methods for Stokes flow: an overview. Comput. Fluids **98**, 221–229 (2014). https://doi.org/10.1016/j.compfluid.2013.11.017
63. Cockburn, B., Singler, J.R., Zhang, Y.: Interpolatory HDG method for parabolic semilinear PDEs. J. Sci. Comput. 1–24
64. Cockburn, B., Zhang, W.: A posteriori error estimates for HDG methods. J. Sci. Comput. **51**(3), 582–607 (2012). https://doi.org/10.1007/s10915-011-9522-2
65. Cockburn, B., Zhang, W.: A posteriori error analysis for hybridizable discontinuous Galerkin methods for second order elliptic problems. SIAM J. Numer. Anal. **51**(1), 676–693 (2013). https://doi.org/10.1137/120866269
66. Dong, B.: Optimally convergent HDG method for third-order Korteweg-de Vries type equations. J. Sci. Comput. **73**(2–3), 712–735 (2017). https://doi.org/10.1007/s10915-017-0437-4
67. Efendiev, Y., Lazarov, R., Moon, M., Shi, K.: A spectral multiscale hybridizable discontinuous Galerkin method for second order elliptic problems. Comput. Methods Appl. Mech. Eng. **292**, 243–256 (2015). https://doi.org/10.1016/j.cma.2014.09.036
68. Efendiev, Y., Lazarov, R., Shi, K.: A multiscale HDG method for second order elliptic equations. Part I. Polynomial and homogenization-based multiscale spaces. SIAM J. Numer. Anal. **53**(1), 342–369 (2015). https://doi.org/10.1137/13094089X
69. Fortin, M.: An analysis of the convergence of mixed finite element methods. RAIRO Anal. Numér. **11**(4), 341–354, iii (1977)
70. Fu, G., Cockburn, B., Stolarski, H.: Analysis of an HDG method for linear elasticity. Int. J. Numer. Methods Eng. **102**(3–4), 551–575 (2015). https://doi.org/10.1002/nme.4781
71. Fu, G., Jin, Y., Qiu, W.: Parameter-free superconvergent H(div)-conforming HDG methods for the Brinkman equations. IMA J. Numer. Anal. **39**(2), 957–982 (2018). https://doi.org/10.1093/imanum/dry001

72. Fu, G., Lehrenfeld, C.: A strongly conservative hybrid DG/mixed FEM for the coupling of Stokes and Darcy flow. J. Sci. Comput. (2018). https://doi.org/10.1007/s10915-018-0691-0
73. Fu, G., Qiu, W., Zhang, W.: An analysis of HDG methods for convection-dominated diffusion problems. ESAIM Math. Model. Numer. Anal. **49**(1), 225–256 (2015). https://doi.org/10.1051/m2an/2014032
74. Fu, Z., Gatica, L.F., Sayas, F.J.: Algorithm 949: MATLAB tools for HDG in three dimensions. ACM Trans. Math. Softw. **41**(3), Art. 20, 21 (2015). https://doi.org/10.1145/2658992
75. Fu, Z., Heuer, N., Sayas, F.J.: A non-symmetric coupling of boundary elements with the hybridizable discontinuous Galerkin method. Comput. Math. Appl. **74**(11), 2752–2768 (2017). https://doi.org/10.1016/j.camwa.2017.08.035
76. Gatica, G.N., Sequeira, F.A.: Analysis of an augmented HDG method for a class of quasi-Newtonian Stokes flows. J. Sci. Comput. **65**(3), 1270–1308 (2015). https://doi.org/10.1007/s10915-015-0008-5
77. Gatica, G.N., Sequeira, F.A.: A priori and a posteriori error analyses of an augmented HDG method for a class of quasi-Newtonian Stokes flows. J. Sci. Comput. **69**(3), 1192–1250 (2016). https://doi.org/10.1007/s10915-016-0233-6
78. Gatica, G.N., Sequeira, F.A.: Analysis of the HDG method for the Stokes-Darcy coupling. Numer. Methods Partial Differ. Equs. **33**(3), 885–917 (2017). https://doi.org/10.1002/num.22128
79. Gopalakrishnan, J., Li, F., Nguyen, N.C., Peraire, J.: Spectral approximations by the HDG method. Math. Comput. **84**(293), 1037–1059 (2015). https://doi.org/10.1090/S0025-5718-2014-02885-8
80. Gopalakrishnan, J., Solano, M., Vargas, F.: Dispersion analysis of HDG methods. J. Sci. Comput. (2018). https://doi.org/10.1007/s10915-018-0781-z
81. Griesmaier, R., Monk, P.: Error analysis for a hybridizable discontinuous Galerkin method for the Helmholtz equation. J. Sci. Comput. **49**(3), 291–310 (2011). https://doi.org/10.1007/s10915-011-9460-z
82. Griesmaier, R., Monk, P.: Discretization of the wave equation using continuous elements in time and a hybridizable discontinuous Galerkin method in space. J. Sci. Comput. **58**(2), 472–498 (2014). https://doi.org/10.1007/s10915-013-9741-9
83. Hungria, A., Prada, D., Sayas, F.J.: HDG methods for elastodynamics. Comput. Math. Appl. **74**(11), 2671–2690 (2017). https://doi.org/10.1016/j.camwa.2017.08.016
84. Kabaria, H., Lew, A.J., Cockburn, B.: A hybridizable discontinuous Galerkin formulation for non-linear elasticity. Comput. Methods Appl. Mech. Eng. **283**, 303–329 (2015). https://doi.org/10.1016/j.cma.2014.08.012
85. Kirby, R.M., Sherwin, S.J., Cockburn, B.: To CG or to HDG: a comparative study. J. Sci. Comput. **51**(1), 183–212 (2012). https://doi.org/10.1007/s10915-011-9501-7
86. Lederer, P.L., Lehrenfeld, C., Schöberl, J.: Hybrid discontinuous Galerkin methods with relaxed H(div)-conformity for incompressible flows. Part I. SIAM J. Numer. Anal. **56**(4), 2070–2094 (2018). https://doi.org/10.1137/17M1138078
87. Lehrenfeld, C.: Hybrid discontinuous Galerkin methods for solving incompressible flow problems. Rheinisch-Westfalischen Technischen Hochschule Aachen (2010)
88. Lehrenfeld, C., Schöberl, J.: High order exactly divergence-free hybrid discontinuous Galerkin methods for unsteady incompressible flows. Comput. Methods Appl. Mech. Eng. **307**, 339–361 (2016). https://doi.org/10.1016/j.cma.2016.04.025
89. Lu, P., Chen, H., Qiu, W.: An absolutely stable *hp*-HDG method for the time-harmonic Maxwell equations with high wave number. Math. Comput. **86**(306), 1553–1577 (2017). https://doi.org/10.1090/mcom/3150
90. Moro, D., Nguyen, N.C., Peraire, J., Gopalakrishnan, J.: A hybridized discontinuous Petrov-Galerkin method for compresible flows. Am. Inst. Aeronaut. Astronaut. (2011). https://doi.org/10.2514/6.2011-197
91. Mustapha, K., Nour, M., Cockburn, B.: Convergence and superconvergence analyses of HDG methods for time fractional diffusion problems. Adv. Comput. Math. **42**(2), 377–393 (2016). https://doi.org/10.1007/s10444-015-9428-x

92. Nédélec, J.C.: Mixed finite elements in R^3. Numer. Math. **35**(3), 315–341 (1980). https://doi.org/10.1007/BF01396415
93. Nédélec, J.C.: A new family of mixed finite elements in R^3. Numer. Math. **50**(1), 57–81 (1986). https://doi.org/10.1007/BF01389668
94. Nguyen, N., Peraire, J., Cockburn, B.: A hybridizable discontinuous Galerkin method for the incompressible Navier-Stokes equations. Am. Inst. Aeronaut. Astronaut. (2010). https://doi.org/10.2514/6.2010-362
95. Nguyen, N.C., Peraire, J., Cockburn, B.: An implicit high-order hybridizable discontinuous Galerkin method for linear convection-diffusion equations. J. Comput. Phys. **228**(9), 3232–3254 (2009). https://doi.org/10.1016/j.jcp.2009.01.030
96. Nguyen, N.C., Peraire, J., Cockburn, B.: A hybridizable discontinuous Galerkin method for Stokes flow. Comput. Methods Appl. Mech. Eng. **199**(9–12), 582–597 (2010). https://doi.org/10.1016/j.cma.2009.10.007
97. Nguyen, N.C., Peraire, J., Cockburn, B.: High-order implicit hybridizable discontinuous Galerkin methods for acoustics and elastodynamics. J. Comput. Phys. **230**(10), 3695–3718 (2011). https://doi.org/10.1016/j.jcp.2011.01.035
98. Nguyen, N.C., Peraire, J., Cockburn, B.: Hybridizable discontinuous Galerkin methods for the time-harmonic Maxwell's equations. J. Comput. Phys. **230**(19), 7151–7175 (2011). https://doi.org/10.1016/j.jcp.2011.05.018
99. Nguyen, N.C., Peraire, J., Cockburn, B.: An implicit high-order hybridizable discontinuous Galerkin method for the incompressible Navier-Stokes equations. J. Comput. Phys. **230**(4), 1147–1170 (2011). https://doi.org/10.1016/j.jcp.2010.10.032
100. Nguyen, N.C., Peraire, J., Reitich, F., Cockburn, B.: A phase-based hybridizable discontinuous Galerkin method for the numerical solution of the Helmholtz equation. J. Comput. Phys. **290**, 318–335 (2015). https://doi.org/10.1016/j.jcp.2015.02.002
101. Oikawa, I.: Hybridized discontinuous Galerkin method for convection-diffusion problems. Jpn. J. Ind. Appl. Math. **31**(2), 335–354 (2014). https://doi.org/10.1007/s13160-014-0137-5
102. Oikawa, I.: A hybridized discontinuous Galerkin method with reduced stabilization. J. Sci. Comput. **65**(1), 327–340 (2015). https://doi.org/10.1007/s10915-014-9962-6
103. Oikawa, I.: Analysis of a reduced-order HDG method for the Stokes equations. J. Sci. Comput. **67**(2), 475–492 (2016). https://doi.org/10.1007/s10915-015-0090-8
104. Oikawa, I.: An HDG method with orthogonal projections in facet integrals. J. Sci. Comput. **76**(2), 1044–1054 (2018). https://doi.org/10.1007/s10915-018-0648-3
105. Peraire, J., Nguyen, N., Cockburn, B.: A hybridizable discontinuous Galerkin method for the compressible Euler and Navier-Stokes equations. Am. Inst. Aeronaut. Astronaut. (2010). https://doi.org/10.2514/6.2010-363
106. Qiu, W., Shen, J., Shi, K.: An HDG method for linear elasticity with strong symmetric stresses. Math. Comput. **87**(309), 69–93 (2018). https://doi.org/10.1090/mcom/3249
107. Qiu, W., Shi, K.: An HDG method for convection diffusion equation. J. Sci. Comput. **66**(1), 346–357 (2016). https://doi.org/10.1007/s10915-015-0024-5
108. Qiu, W., Shi, K.: A superconvergent HDG method for the incompressible Navier-Stokes equations on general polyhedral meshes. IMA J. Numer. Anal. **36**(4), 1943–1967 (2016). https://doi.org/10.1093/imanum/drv067
109. Raviart, P.A., Thomas, J.M.: A mixed finite element method for 2nd order elliptic problems. In: Mathematical aspects of Finite Element Methods (Proc. Conf., Consiglio Naz. delle Ricerche (C.N.R.), Rome, 1975), pp. 292–315. Lecture Notes in Mathematical, vol. 606. Springer, Berlin (1977)
110. Rhebergen, S., Cockburn, B.: A space-time hybridizable discontinuous Galerkin method for incompressible flows on deforming domains. J. Comput. Phys. **231**(11), 4185–4204 (2012). https://doi.org/10.1016/j.jcp.2012.02.011
111. Rhebergen, S., Cockburn, B.: Space-time hybridizable discontinuous Galerkin method for the advection-diffusion equation on moving and deforming meshes. In: The Courant-Friedrichs-Lewy (CFL) Condition, pp. 45–63. Birkhäuser/Springer, New York (2013). https://doi.org/10.1007/978-0-8176-8394-8_4

112. Sánchez, M.A., Ciuca, C., Nguyen, N.C., Peraire, J., Cockburn, B.: Symplectic Hamiltonian HDG methods for wave propagation phenomena. J. Comput. Phys. **350**, 951–973 (2017). https://doi.org/10.1016/j.jcp.2017.09.010
113. Schöberl, J., Lehrenfeld, C.: Domain decomposition preconditioning for high order hybrid discontinuous Galerkin methods on tetrahedral meshes. In: Advanced Finite Element Methods and Applications. Lecture Notes in Applied and Computational Mechanics, vol. 66, pp. 27–56. Springer, Heidelberg (2013). https://doi.org/10.1007/978-3-642-30316-6_2
114. Sevilla, R., Giacomini, M., Karkoulias, A., Huerta, A.: A superconvergent hybridisable discontinuous Galerkin method for linear elasticity. Int. J. Numer. Methods Eng. **116**(2), 91–116 (2018). https://doi.org/10.1002/nme.5916
115. Soon, S.: Hybridizable Discontinuous Galerkin Method for Solid Mechanics. University of Minnesota (2008). https://books.google.com/books?id=0s7KAQAACAAJ
116. Soon, S.C., Cockburn, B., Stolarski, H.K.: A hybridizable discontinuous Galerkin method for linear elasticity. Int. J. Numer. Methods Eng. **80**(8), 1058–1092 (2009). https://doi.org/10.1002/nme.2646
117. Stanglmeier, M., Nguyen, N.C., Peraire, J., Cockburn, B.: An explicit hybridizable discontinuous Galerkin method for the acoustic wave equation. Comput. Methods Appl. Mech. Eng. **300**, 748–769 (2016). https://doi.org/10.1016/j.cma.2015.12.003
118. Stenberg, R.: Postprocessing schemes for some mixed finite elements. RAIRO Modél. Math. Anal. Numér. **25**(1), 151–167 (1991)
119. Yadav, S., Pani, A.K.: Superconvergent discontinuous Galerkin methods for nonlinear parabolic initial and boundary value problems. J. Numer. Math. (2018). https://doi.org/10.1515/jnma-2018-0035
120. Yadav, S., Pani, A.K., Park, E.J.: Superconvergent discontinuous Galerkin methods for nonlinear elliptic equations. Math. Comput. **82**(283), 1297–1335 (2013). https://doi.org/10.1090/S0025-5718-2013-02662-2
121. Yakovlev, S., Moxey, D., Kirby, R.M., Sherwin, S.J.: To CG or to HDG: a comparative study in 3D. J. Sci. Comput. **67**(1), 192–220 (2016). https://doi.org/10.1007/s10915-015-0076-6

Index

S. Du and F.-J. Sayas, *An Invitation to the Theory of the Hybridizable Discontinuous Galerkin Method*, SpringerBriefs in Mathematics,
https://doi.org/10.1007/978-3-030-27230-2

Printed in the United States
By Bookmasters